W9-BAN-998

FINDING HIGHER GROUND

Finding Higher Ground

Adaptation in the Age of Warming

AMY SEIDL

Beacon Press, Boston

Beacon Press
25 Beacon Street
Boston, Massachusetts 02108-2892
www.beacon.org

Beacon Press books
are published under the auspices of
the Unitarian Universalist Association of Congregations.

14 13 12 11   8 7 6 5 4 3 2 1

This book is printed on acid-free paper that
meets the uncoated paper ANSI/NISO specifications
for permanence as revised in 1992.

Text design by Jody Hanson, Wilsted & Taylor Publishing Services

Library of Congress Cataloging-in-Publication Data
Seidl, Amy.
Finding higher ground : adaptation in the age of warming /
Amy Seidl.
p. cm.
Includes bibliographical references.
ISBN 978-0-8070-8598-1 (hardback : acid-free paper)
1. Acclimatization. 2. Plants—Adaptation. 3. Adaptation (Biology)
I. Title.
QH546.S453 2011
578.4'2—dc22   2010050223

To Lawrence E. Seidl

*We see these beautiful co-adaptations most plainly
in the woodpecker and the mistletoe; and only a little
less plainly in the humblest parasite which clings to the
hairs of a quadruped or feathers of a bird; in the structure
of the beetle which dives through the water; in the plumed
seed which is wafted by the gentlest breeze; in short,
we see beautiful adaptations everywhere and
in every part of the organic world.*

—CHARLES DARWIN, *ON THE ORIGIN OF SPECIES*

Contents

Preface

To *adapt* is to become suited to the conditions around you. In the natural world, *adaptation* describes the traits that help organisms cope with the environments they live in. Songbirds migrate between northern and southern hemispheres as an adaptation to seasonal change, and plants produce toxins as an adaptation to deter insects from eating their leaves. Similarly, an animal that exhibits the adaptive coloring of camouflage—a fawn in dappled light, a praying mantis disguised on a blade of grass—is expected to live longer (and presumably have more offspring) than an undefended animal that stands out in its surroundings. If a trait is adaptive, it confers success.

In the human realm, *adaptation* means to adjust, where one's actions are suited to conditions at hand. People without work adapt by spending less money, children who move in the middle of the year adapt by making new friends, and persons who lose their legs adapt by relying on wheelchairs or artificial limbs.

In *Finding Higher Ground,* I use *adaptation* to describe this range of meaning: the biological and ecological responses to changing conditions as well as the adjustments that humans make to accommodate change in their lives. But my principal topic is how human and nonhuman life is adapting to the all-encompassing phenomenon of climate change.

While adaptation is the subject of this book, *carbon mitigation* has framed our approach to climate change for several decades. Carbon mitigation is the reduction of greenhouse gases as well as the sequestration of carbon that has already been emitted. Establishing policy that sets targets for lowering emissions and removing and storing already-released carbon (in croplands, newly planted forests, and empty gas wells) has been our measured, if not entirely successful, way forward. And rightly so, given the effect of a carbonated atmosphere on the long-term stability of Earth's systems.

Mitigation, however, is distinctly different from climatic adaptation, as the latter focuses on how to live through and prepare for the phenomenon itself. How will we, as individuals, communities, and nation-states, anticipate and respond to climate change in our lives? How will we build resilience into our social and physical infrastructure (transportation, energy and food systems, and our homes) to help us recover from its effects and adapt to what lies ahead? As importantly, how can we learn from the ways in which the biological world is already adapting around us?

In 2009 I wrote *Early Spring: An Ecologist and Her Children Wake to a Warming World,* a book that explored the signs of climate change in a New England landscape. Here, I examined the ecological and cultural changes in a world seen close at hand, in places where anyone who observes nature could find them. I described how spring flowers bloom before pollinators arrive, how ponds no longer freeze, and how animal migrations occur at unexpected times.

It was while researching these early signals that I came to think about adaptation. If climate is a major factor that shapes environmental conditions, how is the natural world not only shifting but adapting to climate change? Moreover, how are people adjusting to the same set of phenomena? Clearly we are reacting when severe weather events occur, but how are we responding to the global reality that climate change will define the next thousand years, a millennium of warming that warrants its own geologic age?

In *Finding Higher Ground*, I explore adaptation as it is taking place in these two realms: the ecological and the human. I begin by showing how adaptation differs from carbon mitigation, suggesting that it is time to approach climate change with an adaptive framework as much as a mitigative one. Next, I explore how natural systems are adapting. Research, particularly in North American ecosystems, has shown that climate change is an agent of natural selection: in places, the warming climate has become an evolutionary force. Organisms are changing what they eat and when they migrate. They are reconfiguring their niche space in a biological effort to succeed in a world that is experiencing radical change. The biological clocks of mosquitoes are being reset, and squirrels are birthing earlier to take advantage of new temperature regimes. Each of these examples illustrates how the process of adaptation is occurring as conditions come into flux, the same conditions that humans will need to respond to as well.

In the latter half of *Finding Higher Ground*, I explore how people are adjusting on individual, local, regional, and national scales. I begin by demonstrating that the way we grow food is changing as the climate warms and the weather becomes more variable. Farmers and gardeners are experimenting with new crops, new varieties, and new ways of growing, all in an effort to anticipate (and where possible benefit from) the new conditions at hand.

Next, I examine how energy systems are under a similar pressure to adapt, not only to climate change but to a future without fossil fuels. Here too people are experimenting. Energy technologies that emit little or no carbon—like solar, wind, geothermal, and hydropower—are being perfected, and people concerned about the effects of climate change on their individual well-being are buying them. Like mosquito biology tracking the effects of climate change, individuals are pursuing technologies that gain them success and a measure of security in a warming world.

Enter the concept of *cultural transition*. Transitioning away from a carbonated world is in large part what adapting to climate change means for human populations. As a term, *transition* captivates those who want to move beyond fossil fuels and toward actions and technologies that are sustainable in a broad sense. *Transition* also carries with it the cultural history of humans who have innovated their way through difficult periods before. Whether it was inventing a source of rubber when imports of natural rubber from Asia were barred, or progressing from wagons to railroads as settlers moved west, our history is full of moments of transition. During a time of warming, cultural transitions will happen again. Not only will our physical infrastructure develop toward emitting little or no carbon, it will be influenced by our legacy of carbon use.

Understanding that carbon mitigation will not be enough and that we must wholeheartedly change the way we live, people are looking for philosophical guidance. They are, after all, pioneering utterly different scenarios for the future. Enter the American tradition of pragmatism and the established concepts of self-reliance and sufficiency. These views give people reason to enact practical measures to a problem whose scale is at times overwhelming. Home gardens, home-based renewable-power systems, and nineteenth- and twenty-first-century technologies like clotheslines and fuel cells, respectively, are redefining how people choose to live given the state of the world and their complicity in what it has become.

Finding Higher Ground is a hopeful book, because it not only tells how adaptation is emerging, it confronts the forecast of collapse. It is true: we face a turbulent future. There is no doubt that there will be tremendous species loss, human suffering, and conflict that arises from compromised landscapes. Scientists tell us that the world will change beyond what most of us can comprehend. But sometimes, advocates' emphasis on urgency, meant to promote action, comes across as fatalistic, envisioning an unwinnable scenario that instead induces paralysis.

In my life, endeavors to adapt to a warming world move me from despondency to motion. Like the disaster-readiness supplies I keep in the crawl space of my home—spare batteries, hand-crank flashlight, water and food for two weeks—responding to global warming with practical actions revises my thinking about what lies ahead. I feel less vulnerable because the preemptive measures I take are not only empowering, they encourage me to belong to the future. In essence, articulating a confidence in our ability to adapt to climate change is a claim for persistence.

We are at a turning point. Realizing that our carbon-infused culture, economy, and lifestyle endanger human and nonhuman life, a transition to new ways of being is prescribed. While mitigating climate change is essential, adapting to and through centuries of warming is paramount. Fortunately, adaptive designs and practices can be informed by the rich history of life on Earth as well as by contemporary ecological and evolutionary responses found in nature. In *Finding Higher Ground* the stories of animals, plants, and people adapting to a warming world express trust in our ability to adjust to changing conditions, even radical ones, and to establish a voice for resilience in uncertain times.

CHAPTER 1

Adapting to a Carbonated World

Life on the planet has entered a new age, the *Age of Warming*. Glaciers are retreating, leaving bare ground behind them, and historic weather records are broken left and right. Like the Hypsithermal, the time seven thousand years ago when Earth warmed suddenly, ours is a new time.

In the past, change of this magnitude was caused by volcanic eruptions, cyclic astronomical events, or random ones like asteroid collisions. Now, because of the fact that most of the world's economies are powered with fossil fuels, and our ability to exploit and consume all manner of natural resources, we have garnered the distinction of planetary engineers, and an epoch all our own—the Anthropocene.

When climate scientists first recognized that Earth's atmosphere was warming, they asked: what effect will it have on nature? This was a seemingly obvious question. In actuality, understanding the range of effects is extremely complicated, and researchers quickly realized how far-reaching those effects on living systems would be. At first, getting at the questions experimentally meant using blunt techniques like hanging electric heat lamps above a mountain meadow to raise the air temperature and measuring the response of wildflowers (when they leafed out, flowered, and set seed) growing nearby. In time, the experiments became far more sophisticated. One approach was called free-air carbon dioxide enrichment (FACE). FACE experiments were designed around an outdoor chamber that maintained a steady input of carbon dioxide in otherwise changing ambient conditions: wind speeds and direction varied, rain could fall, but all the while carbon dioxide remained elevated. Researchers regularly chose 500 parts per million (ppm) to equate to a "carbonated atmosphere." Why? Because 500 ppm represented an approximate doubling from preindustrial carbon dioxide levels (280 ppm times two), an emissions trajectory the planet was demonstrably on.

These were the first attempts to understand the consequences of global warming for the natural world. While entirely simplified, the experiments established upper thresholds for levels that plants could withstand, and teased out whether plants ever benefited or were always negatively affected by elevated carbon. More importantly, they produced data for the larger-scale modeling experiments that addressed whole-system processes. As empirical data became available and more studies focused on ecosystems, it became clear that political efforts should be made to keep atmospheric concentration below 500 ppm. Climate models that used an approximate doubling of carbon resulted in an ice-free planet with constantly changing shorelines and erratic weather conditions. In 2008 James Hansen, chief climate scientist at the National Atmospheric and Space Administration (NASA), soberly concluded that a doubling of carbon dioxide is "a figure beyond which life on Earth is adapted."

On many levels, a highly carbonated atmosphere is already a foregone conclusion. While the Age of Warming is not irreversible, it cannot be reversed for many centuries, owing to the longevity of the molecules in the atmosphere and to the inertia in the carbon cycle itself; it takes time for carbon dioxide and other greenhouse gases to degrade into nonwarming gases or to be taken up by carbon sinks such as the oceans, forests, and croplands. This idea presents a quandary for the climate movement. On the one hand, a full-court press toward carbon mitigation so that the atmosphere stabilizes at 350 ppm—the concentration of carbon dioxide that many scientists agree should be our target—is absolutely necessary. On the other hand, given the fact that the atmosphere is already beyond 380 ppm and climbing, and that climate change is projected to continue well into the next millennium, it behooves us to promote adaptation alongside carbon mitigation.

There are thresholds, however, beyond which life's ability to adapt will be seriously challenged. At one point, climate scientists thought that an atmospheric concentration of 450 ppm was an appropriate target for governments and climate negotiators. Now we realize that 450 ppm, which would result in an average temperature increase of 3.6 degrees Fahrenheit (2 degrees Celsius), makes for a tenuous world, one even the most optimistic of technologists are challenged to envision. Of this level, James Hansen comments, "If you leave us at 450 ppm for long enough it will probably melt all the ice—that's a sea rise of 75 meters. What we have found is that the

target . . . is a disaster—a guaranteed disaster." For instance, at 450 ppm Vermont's climate will resemble At-
lanta, Georgia, with an average temperature in January
of 40 degrees Fahrenheit (4 degrees Celsius).
This kind of radical change in the world's climate is
just one of numerous predictions that force us to envision
new infrastructures. Bridges will need to be built differently to withstand more frequent flooding. In coastal
towns, seawalls will be needed to protect against higher
swells, more frequent storms, and sea-level rise. For instance, on March 1, 2010, powerful waves combined
with strong winds hit the coast of southwestern France
in the early hours before dawn. Waves broke through
aged and unsound seawalls, inundating the villages of
L'Aiguillon-Sur-Mer and La Faute-Sur-Mer. Fifty-two
people were killed while they slept.

France is a coastal country. It has 6,200 miles of seawalls, many of which are in disrepair and could not have
withstood the late winter storm. How many other coastal
towns are similarly vulnerable?

It is said that mitigation moves at the speed of politics while adaptation moves at the speed of events. Carbon mitigation will not stave off storm surges. Like the
double-stranded curve of DNA, adaptation is mitigation's yoked second strand. Intertwined, they present the
path forward. Adaptation will be most effective at local
and regional scales; the French will build their seawalls
higher, just as the people of Louisiana will build their
levees stronger and higher as a result of the devastation
wrought by Hurricane Katrina. Communities, and the
people who live in them, feel the effects of a changing
climate. Therefore, adapting to climate change, and
the host of environmental, economic, and social prob-

lems that come with it, will take place in situ. Similar to the ways firehouses exist for the people in proximity to them, response to climate change when the localized crises descend happens where we live.

For developed economies such as the United States, adaptation and mitigation can go hand in glove. We are better positioned—geographically, economically, and politically—to adapt to climate change and combine the strategies of reduction with adaptation. Placing micro-hydro turbines—bread-loaf-sized engines whose blades spin when running water hits them—in small streams will generate carbon-free power in regions where more rain is a result of global warming. Similarly, producing local food strengthens food security in an era of more variable and extreme weather.

Nonindustrialized countries, especially in the southern hemisphere, have little national carbon to mitigate; they just don't emit much. The 139 countries that formed the group of least developed nations at the climate talks in Copenhagen, Denmark, in 2009 contribute less than 10 percent to total world emissions. Some, like the island state of Tuvalu, contribute essentially nothing. Yet these countries are the ones most impacted by climate change. For them, adaptation, and in some instances survival, dominates their response to a changing planet, and their sovereignty depends on mitigation by other nations. The conservation biologist David Orr likens proponents of mitigation to individuals who choose to "turn the water off in an overflowing tub before mopping." While that seems to make sense, proponents of adaptation from the least developed nations realize they have little access to the spigot and may have to evacuate the flooded house altogether.

Enter the term *resilience,* the capacity to absorb and adapt to system shocks. In climate-change parlance, it is the ability to recover after being hit by a weather event linked to warming: a hurricane, heat wave, or seven inches of rain in three days. Depending on the magnitude of the shock, a community's resilience can be severely tested. In April 2010 Rio de Janeiro experienced more rain in one day than usually falls in the entire month. Houses collapsed, rivers were overrun, and water tanks burst. David Zee, a professor of oceanography at the University of the State of Rio de Janeiro, attributed the severity of the storm to "global climate changes that have local effects." For Rio, the combination of an increase in El Niños, warmer surface waters in the Pacific, and deforestation of the Amazon has led to radical changes in the hydrologic cycle. Like detecting the multiple factors that affect a person's cancer risk, detecting which factors are driving radical weather demands a kind of forensics unknown to meteorology and climate science.

Coral reefs have become the poster child for extinction in oceanic environments. Bleached and dying reefs—their polyps devoid of the brilliant colors of clown fish and sea anemone—are common images in mainstream media. Like other ecosystems in peril, reefs are under simultaneous and multiple threats: pollution, sedimentation from runoff, warmer ocean temperatures, ocean acidification from high concentrations of absorbed carbon dioxide, and overharvesting of the coral fish that dwell in them, among others. But off the coast of Tanzania, in the triangle of ocean that includes the island of Madagascar and the Kenyan coast, there exist coral reefs with unusual resilience. Termed "super reefs," these communities are more able than others to recover from disturbance. Tim McClanahan, a marine ecologist with the Wildlife Conservation Society, asserts that the reefs' exposure to unusual variations in temperature and currents has been the selective pressure responsible for their resilience. This, in addition to the high diversity of species that naturally exists on African reefs, allows the super reefs to respond quickly to bleaching events and to recolonize afterward.

Terms like *resilience* and *vulnerability* apply not only to nation-states and natural communities but also to individuals. Personally, each of us knows what vulnerability feels like, when we have been without cover, insecure, or exposed to danger. Moreover, we extend these emotions empathetically and sometimes sympathetically to people exposed to disaster. We wonder if the increased frequency of natural disasters won't affect us too. At the same time, we know what it means to be resilient, to adapt to change, and to respond in a flexible way when our own personal worlds are shocked. We are all, consciously and unconsciously, working to reduce our vulnerability.

Striving for resilience in the Age of Warming is becoming a priority for all regions of the world. Reports by the economist Nicholas Stern and others urge proactive adaptation, in addition to mitigation, as a far less expensive approach than reacting to crises. As importantly, proactive adaptation instigates long-time-horizon thinking. It forces us to plan for the effects of climate change on our communities not ten years out but thirty, fifty, even one hundred years from now.

While climate change will be a problem that spans generations, its ultimate resolution—a decarbonated atmosphere—will take place in the future. This dynamic is very much like the conservation of wildlands. Knowing that species extinction is accelerating, we create national parks, refuges, and preserves to benefit generations of human and nonhuman life yet to be born. Similarly, our actions to mitigate and adapt to climate change are actions taken on behalf of the near future—the one in which we may be a participant—and the future-future— in which we won't.

"It is not generally appreciated that atmospheric temperature increases . . . are not expected to decrease significantly even if carbon emissions cease," wrote Susan Solomon, a climate scientist and leading author with the Intergovernmental Panel on Climate Change (IPCC). The carbon cycle, like other nutrient cycles, moves carbon through the atmosphere, biosphere, and oceans according to strict physical and chemical laws. In particular, carbon is never created or destroyed but cycles infinitely through various carbon reservoirs. These reservoirs include the oceans, where carbon dioxide mixes with carbonate ions such as calcium carbonate (the key ingredient in seashells) to form carbonic acid.

Plants also take up carbon during photosynthesis and store it as biomass, either briefly, as in algae, or for centuries, as in the bristlecone pine. And then there is the atmosphere, a reservoir of carbon-based gases that reradiate Earth's heat energy, often for many decades, before degrading. Solomon writes, in the scientific journal *Proceedings of the National Academy of Sciences*, about the long-tail effects of reducing carbon emissions, and warns that climate change is a thousand-year phenomenon, in part because of the longevity of these greenhouse gases. Carbon comes apart from oxygen in the carbon dioxide molecule, and hydrogen is cleaved from carbon in the methane molecule, but only after one hundred years and ten years, respectively.

The longevity of greenhouse gases, coupled with inertia in the exchange of carbon among reservoirs—atmosphere, ocean, biosphere—is what is responsible for the long-term nature of climate change. Even with rapid mitigation, the legacy of a warming world will remain. "Irreversible is defined here as a time scale exceeding the end of the millennium in year 3000," Solomon writes. This is the future-future, as distanced a perspective as most of us are ever likely to conceive.

The near irreversibility of climate change forces us to confront a time unlike any period in the past one hundred thousand years of human history. A melting Arctic and changing seasons are signals of this warming world. They stretch our sense of time and challenge our sense of equity between generations and people from different parts of the world. The irreversibility of climate change insists that we reconcile ourselves to the real limits of the planet, and how hard and fast the laws of chemistry and physics are.

This understanding is akin to what people face when their lives become desperate as a result of reckless and self-destructive behavior. It might come in the form of an accident, overdose, or threat of dire consequence, from a doctor or a judge. If we find ourselves in this despairing place, most of us submit that our health is of paramount importance. We give in to treatment, rest, or recovery, and with time—and often a hellish transition—we are surprised by the clarity that comes postreconciliation; we can see which paths lead to health and which lead to mortal danger. But it is the clarity of the two choices that is at the heart of my metaphor, the crystalline understanding for those who have hit rock bottom: from this point forward, things must change.

For many climate activists, our low point was reached at the international climate talks in Copenhagen in 2009. Going into the talks, analysts predicted future atmospheric carbon based on the reduction goals of the 192 countries that participated: 32 percent below 1990 levels for the United Kingdom; 5 percent for Australia; 4 percent for the United States; and 100 percent for Tuvalu were among them. Based on these goals, the projection was 770 ppm, far and away beyond the limit of atmospheric carbon necessary to stabilize Earth's climate. In the eyes of the activists, it was our collective rock bottom.

Rock bottom is something other human civilizations have had to face. The story of the Norse in Greenland is an iconic example of the effect of climate on settlement success. Erik the Red first explored Greenland during favorable conditions in the tenth century, a serendipitous event for a Norse raider who was searching for bounty and, incidentally, fleeing a criminal past. Abundant fish, sea mammals, herds of caribou, and rich green meadows were the selling points he used to form a Danish colony far distant from the motherland. Advertising the new territory with the attractive name of Greenland, the Norse adventurer returned to the Arctic island in 986 AD with twenty-five ships crowded with colonists.

The Norse established settlements on the southern and western coasts, on land the Thule-Inuit people called Kalaallit Nunaat, meaning Land of the Kalaallit. As many as seven thousand Europeans lived on Greenland during its heyday, and for approximately five hundred years they extracted a living from the rocky soils. Ar-

chaeological records, derived from Norse sagas and gar-
bage middens, describe a pastoral subsistence composed
of medieval dairies, migratory seal and cod hunting, and
the export of walrus and narwhal ivory, which were used
to make valuable Catholic icons.

After the period during which Erik the Red founded
the Danish colonies, the northern hemisphere entered
the Little Ice Age. Freezing late-spring temperatures,
advancing glaciers, harsh weather throughout the year,
and pack ice that invaded the southern coastlines well
into the previously warm months of May and June began
to characterize the climate. It was a chilling time and
it extended across northern Europe. Historical records
from Poland and Russia depict a deep famine in which
the desperate foraged on pine bark and parents inden-
tured those children they were unable to feed.

The Norse people living in Greenland were similarly
wretched. Pack ice kept ships from accessing the Arctic island and Norse villages, breaking contact with the
motherland and the relief that came with it. Yet others,
living within reach of the Norse on the same island,
fared well during this time, even thrived. These were
ancestors of the Inuit, Paleo-Eskimos who paddled seal-
skin kayaks and used toggle-head harpoons to hunt seals
throughout the year. Unlike the Norse, who suffered
until desertion, the Inuit lived successfully through the
Little Ice Age. Their middens—pored over by the same
archaeologists who mined the Norse trash heaps for
clues of demise—depict plenty, with the marrow still in
the discarded bones.

When we think about adapting to climate change,
many writers, including Jared Diamond in his book
Collapse, point to the failure of societies that refused

to adapt or that didn't adhere to the signs of change around them. Easter Islanders, the Maya, and the Norse of Greenland are the iconic cases. But the Inuit remain on Greenland to this day; six hundred years after the Norse perished with the bodies of their dairy cows and treasured hunting dogs beside them. Clearly, the indigenous inhabitants were better adapted than their counterparts when conditions changed. This isn't to say that climate change was the sole reason for the demise of the Norse. Economic marginalization and a declining population in a society dependent on communal labor were also factors. Still, the Inuit adapted to the glacial temperatures that took hold of the region, and evidence shows that their populations increased on Greenland during this time. Their tools—the toggle-head harpoon and seal-skin kayak—were appropriate technologies, while the Norse's nets and reliance on migratory game left fatal gaps in their subsistence, ones that widened when their seasonal pastoralism was thwarted by late springs, glacial advance, and eroding soils.

"Had the [Norse] adopted toggling harpoons and other traditional ice-hunting technologies from their Inuit neighbors, they could have taken ring seal year round and avoided the late spring crises," writes Brian Fagan in *The Little Ice Age,* a book that examines the role of climate in human history. Fagan theorizes that an aversion to Inuit paganism from the staunchly Catholic Norse, as well as ideologies grounded in European culture, prohibited them from adopting, and thereby adapting to, the successful lifestyle of the Inuit.

The Norse people were in Greenland for approximately five hundred years, twelve generations of opportunity to blend native practices with European ones, twelve moments to fit themselves to local conditions. While the Norse did adopt some Inuit customs—they tracked snowshoe hares and ptarmigans, hunted communally, and fished for cod—in the end they were unable to evolve culturally. Their resistance to change is recurrent in human history. In an era of climate change the question surfaces again: Will we resist? Or can we adapt to radically shifting conditions? What is getting in our way?

~~~~~

Deirdre Heekin turns heads. Her streaked white-blond hair and stylish eyeglasses give her an urban air, but she's as often in a pair of rubber Wellington boots as in high heels. Deirdre is a gardener, restaurateur, and winemaker, an all-around food connoisseur. As a viticulturist, Heekin and her partner, Caleb Barber, grow grapes and berries that they turn into local wines and cordials, unique spirits that diners taste at their nine-table bistro in Woodstock, Vermont. Alert to the changes in regional weather and climate, Heekin and Barber are testing varieties to determine which will do well in variable weather, higher temperatures, and greater precipitation.

"I'm experimenting with grapes that thrive in warmer climates," Heekin tells me after we meet in a bookstore where her latest offering, *Libation: A Bitter Alchemy*, is being sold. By grafting warm-loving European grapes such as Zweigelt, Marquette, and La Crescent onto native stock, she is hoping to establish lines that are both hardy with respect to changing conditions and tasty to drink.

Heekin probes the genetics of plants the way painters

create their palettes—a little bit of this and a little bit of that. Heekin's imagination allows her to conceive of the effects of climate change in the realm of the age-old service of providing food and drink. She is designing strategies to steer her through the radical changes ahead. She travels to France, Italy, and Canada, where grapes grow under temperature regimes that she expects to see on her own farm. She's taking notes, using science and intuition alike.

But she isn't entirely optimistic. Heekin acknowledges her sadness for what will be lost; she's already pining for the plants she will no longer be able to grow: fall peas and spring turnips. But as a mother yearns for the infant when her daughter grows into a girl, or the way a son remembers his father's sharp focus after he is lost to dementia, we find a way forward. "We can adapt ourselves," Heekin says, regaining her sanguine perspective, "not only as cultivators choosing varieties that will do well—for their own good as well as ours—but in the broader sense. We can shift over to new ways of being."

How to shift, mitigate, and adapt. This is the question each individual, community, and nation is facing. Some people who accept climate change are beginning with modest changes: replacing incandescent lightbulbs with compact fluorescents, turning down the thermostat, taking the bus, forgoing meat. For communities there are park-and-ride stations, tree planting, and the insulating of municipal offices. There is low-hanging fruit at the national level too: provide tax breaks for efficient appliances and solar hot water, fund mass transit and high-speed rail, and green the roofs of federal buildings. These actions represent the first phase of transition to a low-carbon world. But as the transition unfolds, broader and deeper changes must take place. The challenge we face is how to enact strategies that are mitigative, adaptive, and based in ecological goals beyond climate change.

What might these look like? "No-regrets strategies" yield benefits to society in spite of climate change and are the least sensitive to uncertainty; they result in positive change in any case. An example is tackling inefficiencies across our physical infrastructure, everything from water leaks to energy-hogging appliances to waste heat. A second no-regrets strategy is the restoration of riparian areas. Wetlands, in addition to filtering toxins, supplying freshwater to communities, and providing habitat, absorb water resulting from extreme weather. This capacity must be maintained. Adapting to climate change can advance environmental sustainability, and is an acknowledgment that the world has finite limits.

Another set of adaptation strategies are ones that are flexible and reversible. For instance, by restricting new homebuilding from floodplains we protect communities against the disastrous effects of flooding. In the event that we find flooding less problematic than expected, or new technologies arise that protect us from it, urbanization can then proceed.

Finally, incorporating extra safety margins from the outset is also good strategy. Why build a seawall to withstand the frequency and intensity of past hurricanes? Better to build them higher in case the upper end of hurricane intensity occurs. This incurs relatively little extra cost, in contrast to building seawalls anew once they are destroyed. Similarly, we can apply a safety margin in our planning, and challenge communities to think not five years ahead but thirty to fifty years, the minimum needed to reckon with climate change. The Age of

Warming demands that we evolve new ways to imagine the future. By promoting durable strategies in planning and construction that convey the paramount purpose of human persistence, we prepare ourselves for the disruption to come.

~~~~~

I received a postcard in the mail that reads "Segovia, Spain" and depicts a blue-black sky with a full moon rising. There is a familiar script on the back relaying how fine the weather has been, how gracious the hotel, and how delightful the combination of aged Manchego cheese and rioja wine. The postage stamp in the upper right-hand corner has a polar bear rendered in cubist style, and the customary "wish you were here" language closes the card.

But it is the stone aqueduct pictured below the evening sky that draws my attention. It dates from Roman times, well before the *Mayflower* landed at Plymouth, even before the Norse colonized Greenland. For two thousand years the aqueduct has carried water from the Fuente Fría River in the mountains above Segovia to the fertile crop and pastureland that surrounds the city. For twenty centuries water has flowed across the expertly placed stone structure. I find it almost beyond comprehension that no mortar was used in its construction. Rather, chiseled granite blocks were placed together with scientific precision. Some were curved into arcs, others were stacked one on top of another as straight as a carpenter's chalk line. They built with their bare hands what we rely on mechanization to do for us. The aqueduct was used for generations, becoming a fixture in the landscape, like the olive orchards that were nourished by the untold gallons of water conveyed to them. I wonder what similarly useful and enduring infrastructures we can create in this time of warming. I wonder what Roman aqueducts can teach us about adaptation and persistence.

CHAPTER 2

Fitting In

In early spring, when the soil can be worked in my Vermont garden, I plant mizuna, a spicy salad green that germinates well in cool conditions. Reddish grains, the size of poppy seeds, go easily into the soil and disappear into the brown humus of my raised bed. In one week's time, lime-green sprouts contrast with the dark earth around them, and in two week's time I thin the rows, casting out the less robust plants to provide space for the selected few in between. In a month I clip the fernlike leaves of mizuna with a pair of children's scissors we keep for harvesting delicate lettuce and herbs. In time, leafy arugula will grow alongside the earlier mizuna, and sturdy spinach too. These first greens, collected from the garden before apple blossoms attract the first bumblebees and tulips unfurl, commence the ritual of eating from the garden.

The mizuna seeds I plant come from seed companies where varieties are selected for their desirable garden traits: slow to bolt, fast growing, large-leaved, and bitter tasting among them. Originally from wild stock, these seeds do not grow wild plants. They have been domesticated and raised in benign conditions, shielded from the forces of natural selection that run unbridled in nature. They have not had to adapt to the contemporary vagaries of weather, nor have they evolved to compete with neighboring species. Still, a kind of artificial selection has acted in the hand of the horticulturalist who, in greenhouses or experimental garden rows, favored one plant over another, selecting traits the way a dressmaker selects fabrics, determining what best suits the gardener's palette, taste, and eye. It may come as a surprise then that mizuna, a household herb originally from Asia, one that arrived in the United States three hundred years ago and became a domesticated garden green now savored in restaurants across the country, is our best illustration yet of how plants could adapt to climate change.

Drought is not a new condition for the mountains and agricultural plains that extend outward from Los Angeles, where wild mizuna grows. For more than a century, irrigation has supplied water to California's Central Valley, where 90 percent of the vegetables that Americans eat are grown. Still, people who live in Southern California categorize the past as "predrought," reserving the term *drought* for the recent weather that makes other years pale in comparison. It is under these current dry conditions—barely ten years' worth—to which wild mizuna populations have adapted and genetically evolved to track this new set of local conditions.

In the area outside Los Angeles, drought and climate change go hand in hand; the warmer the planet gets, the more drought Southern California experiences. More droughts lead to more wildfires, which lead, as burning biomass releases carbon into the sky, to more warming and more droughts—a positive-feedback cycle that in a time of warming has no end. As I write this the 2009 Station Fire, which will ultimately consume over 160,000 acres north of Los Angeles, is raging. People as far away as Denver, a distance of one thousand miles, complain of the acrid scent of smoky air.

In 1997 Arthur Weis, a biologist at the University of California at Irvine, was studying mizuna. Weis was interested in the genetics of engineered plants, especially rapeseed, a plant farmers grow for oilseed production, and a key ingredient of biofuels. In particular, he wanted to know whether engineered genes could "leak" into unengineered plants, in other words, whether the pollen or seed from engineered plants in cultivated fields can be spread through wild populations via bee pollination, small-mammal movements, or the wind. Weis's interest led him to mizuna, a plant closely related to rapeseed and in the same family, Brassicaceae.

To carry out his lab experiments, Weis collected mizuna seeds from populations that had "escaped" farms and were growing wild in marginal lands. This research led him to understand the genome of mizuna. When the experiment was over, Weis stored the extra seeds he'd collected in a refrigerator, where they became dormant. He had no immediate plans to use them.

Seven years later, California's drought was well under way. Now Weis was interested in whether plants were evolving to survive drought conditions. He set about to develop something he later termed "the resurrection protocol." In Weis's case he didn't have to roll a stone away from a cave; he would open the refrigerator and expose his dormant mizuna seeds to warm soil and sunlight once again. Weis's experiment involved growing the stored seeds from his previous experiment, his "ancestral" seeds, alongside recently harvested "descendant" seeds, ones he'd collected from mizuna plants growing in current dry conditions. Seven years and seven generations separated the two types of seed. Aware that climate conditions had shifted for the mizuna growing near Irvine, California, where Weis and his colleagues worked, and knowing they had ancestral (stored) and descendant (recent) seed available to them, they set about growing both seed types in a common garden.

Common-garden experiments are foundational to the science of evolution. By raising plants from different locales under the same conditions—a common garden—researchers can control factors that otherwise could affect the outcome of the experiment. Slope, exposure, and soil type can confound the results of experiments conducted in the field, biasing results and making it difficult to decipher whether it was something about the location or how the plants were treated that caused the outcome.

Perhaps the first person to use common-garden experiments was Gregor Mendel, a famous Augustinian monk who studied the inheritance of traits in peas. Mendel, a contemporary of Charles Darwin but unknown to him, observed that plant traits vary among individuals. By growing his experimental plants under identical conditions in a common garden that surrounded his abbey, he discovered the importance of inheritance of parental characteristics: parents with smooth-skinned seeds, for example, gave rise to offspring with smooth-skinned seeds.

Environmental conditions at the local or micro scale

can affect plant success. We gardeners know this intuitively when we see plants grow differently from one garden to another. My neighbor and I often share seeds. I'll buy a pack of Russian kale and divide the thimble-sized volume with her. This way our seed supply remains fresh. We are not shelving half-empty packets for a year and risking their viability the following season. But barely a quarter of a mile apart, living on the same hillside, my neighbor's kale plants do better than mine. Is it the compost dressing she puts on her garden that makes her kale so vigorous? Or is it the clumps of acid-loving moss in my own beds that make my kale anemic? Temperature and rainfall are the same on our shared hillside, and we assume that any variation in the seed packet is divided evenly between us. There must be other factors at play.

For the better part of two decades, ecologists have found that plants are responding phenologically to Earth's warming, meaning that they are shifting their seasonal timing. They often produce leaves and flowers earlier to remain aligned with earlier springs. The early onset of life-history events is true for organisms besides plants and include when they mate, have offspring, set fruit, produce seeds, even when they age and die. Birds, mammals, amphibians, and insects are all responding to the early arrival of spring and the lateness of fall in the northern hemisphere, changes that unambiguously mark the existence of global warming.

Still, there are a limited number of ways that species can adjust to climate change. First, they can migrate or move permanently to more suitable conditions; many species are moving up in elevation and poleward in latitude in search of climates they remain suited to. Second, they can acclimate in situ, exhibiting an inherent and flexible response to the conditions where they live. Third, species can evolve adaptive traits and undergo genetic changes that better suit the changing climate. Finally, species that are unable to adjust through migration, acclimatization, or genetic change will decline and many will go extinct.

One of the great ecological and evolutionary questions to emerge from climate change research is this: when do species acclimatize without genetic change and when does climate act as a selective force, favoring traits suited to local conditions? While the outcome may be similar—species are able to survive in novel conditions— the mechanism of achieving survival is very different. Further, both acclimatization and genetic change can take place simultaneously, obliging researchers to tease apart the explanatory power of each mechanism.

By analogy, people who climb the highest peaks in the world (Mount Everest, for instance) acclimate to high elevations by slowly climbing from base camp upwards, gradually increasing the oxygen in their blood. Their bodies respond by making more red blood cells that produce oxygen. This is what is known as a "plastic response," a physiological process that is triggered when local conditions change. Alternatively, Nepalese Sherpas who are born at high elevation have evolved barrel chests and larger lungs. It is presumed that natural selection has favored genetic traits that produce the Sherpa's unique anatomy because it is so well suited to the chronic hypoxia (a deficiency of oxygen to tissues) that the Nepalese endure living as they do above three thousand meters.

Phenotypic plasticity is "a universal property of living things," writes the evolutionary biologist Mary Jane West-Eberhard. In essence, it is the responsiveness of an organism to its environment without the benefit of inheritance. While biologists debate whether plasticity can evolve, some suggest that there might be plasticity genes that confer flexibility *across* an organism's genome. Distinguishing between a plastic response and a genetic, evolutionary change is where much of the research on how organisms are responding to climate lies. It is also the reason why the result from Arthur Weis's common-garden experiment with mizuna is so compelling.

Weis's observation was this: extreme drought in California had altered the growing season, effectively shortening it. Therefore, plants generally (and mizuna specifically, given what he knew of its biology) might be adapting to increasingly dry conditions. With his resurrection protocol in place, Weis asked: Have the descendant mizuna plants adapted to drought during the seven years since he last collected wild seed? He hypothesized that mizuna could respond in one of two ways: plants could exhibit phenotypic plasticity (by using water more efficiently, for instance), or, natural selection mediated by drought could have forced genetic change, favoring individuals that alter the time when they grow, flower, and set seed to correspond with the drier conditions.

Ancestral and descendant mizuna were grown in a common garden: same fertile soil, same light, and same spacing around seeds. Knowing that mizuna can grow in both wet and dry areas, and having seed from both types of sites, Weis watered according to normal pre-drought conditions or current drought conditions. This allowed him to look for differences between the sets of ancestral seeds and descendant seeds as well as between

ancestral seeds and descendant seeds that originated
from a wet or dry site.

What Weis found is what biologists have long con-
templated ever since the concept of evolution, and the
time frame it occurs along, was introduced: evolution
can happen quickly. Not only is evolutionary change
something we can observe in a human lifetime or over
vast time frames. It can occur, as Weis documented, in
seven generations.

Weis found that in less than a decade mizuna plants
whose ancestors had survived extreme drought evolved.
They developed novel strategies to survive with limited
water. They allocated resources differently and flowered
earlier. They "hedged their bets," didn't grow as tall and
didn't produce as many seeds as their ancestors. But they
did reproduce, generating seed before lethal conditions
set in. Theoretically, this strategy is in keeping with the
broader field of plant physiology. By flowering earlier,
plants avoid the worst of the drought, and benefit from
the predictably wet conditions that come in early spring.
Earlier flowering and fewer seeds is what Weis calls the
plant's "drought escape" strategy.

As expected, all the mizuna plants in Weis's experiment survived well under the wet, long-season conditions. If there was plenty of water, it didn't matter
whether the seeds came from wet or dry sites. Sufficient
water meant success for all plants. On the other hand,
survival of plants grown under drought-mimicking,
short-season conditions differed significantly between
plants from wet and dry sites: 90 percent of dry plants
but only 62 percent of wet plants survived drought conditions. The conclusion: plants from dry sites, growing
increasingly droughty with climate change, had become

locally adapted. Alternatively, plants in wet sites were more likely to die because they had not been under the selective pressure to physiologically adapt to dry conditions. In time, as even wet conditions became dry, these individuals would be under selection as well.

Natural populations are acclimating to the "new weather." Some, like mizuna, are changing genetically. Others are exhibiting plasticity. A plant's plastic response to drought is less an on/off switch than a gradation of responses, given conditions. By being flexible, an individual can exhibit characteristics unique to its genetic makeup and the environment it encounters, what is known as a phenotype.

Phenotypic plasticity helps us understand how plants and animals will adapt to climate change. Since 1970 ambient temperatures have risen approximately 1 degree Fahrenheit (0.75 degrees Celsius), yet this may be the least of our worries. Ambient temperatures will increase several times that by 2100: 4 to 7 degrees Fahrenheit (2 to 5 degrees Celsius). In addition to temperature, many more environmental parameters will change as well. Altered precipitation regimes, increased storm activity, and overall changes in seasonality will continue. There is also speculation that volcanic eruptions and earthquakes will increase as glaciers recede and the weight of ice is lifted from Arctic and high-elevation zones. With all this physical change occurring on Earth, how much acclimatization to a warming climate can come from an organism's plasticity?

If one assumes that plasticity is derived from the variable conditions experienced over time, then it is reasonable to assume that Earth's flora and fauna have experienced changes like these in their evolutionary past.

But when the changes become radically different, like
when the Arctic loses its ice or the Antarctic is colonized
by South American flora, then it is less plausible that individuals "carry" that much plasticity with them. While
species extinction is a seriously distressing outcome of
global warming, persistence and adaptation will also be
outcomes. The world is becoming a different place in the
Age of Warming, a dynamic world that is less predict-
able than modern humans have experienced in our time
on Earth.

Peer into the natural world, one close at hand. Perhaps it is a city park whose paths are lined with oak or
maple trees planted in the nineteenth century. Or maybe
you are fortunate enough to walk in a remnant prairie
with freshwater kettle ponds and migratory ducks, or
an old-growth forest with trees whose gigantic trunks
and canopies house thousands of species. Maybe you are
walking in your own backyard, traversing an enclosed
space that you've filled with daylilies, climbing roses, and
garden beds filled with vegetables. All these places—the
ones intended as sanctuary or refuge, the ones cultivated
by gardeners, the wild places with no cultivators or patrons—all are experiencing the agitation of change.

Climate change is affecting every living thing, and
all of life's interactions, too. What we have launched
is colossal. Appreciating what an age of warming will
look like is an enormous challenge to our imagination.
In this tension of knowing and anticipating the future,
we are full of fear and wonder. Our approach to the biological changes we will witness could be duly rational,
even dispassionate, like watching a volcano erupt and
being amazed at the heat and gas released or the way
it radically alters a landscape, forming new islands in

a single moment. Alternatively, being aware of climate disruption forces us to acknowledge that human action has become a selective pressure that is changing the very ecosystems we depend on for our existence. This tension colors our awe, as it should, tainting it with grief while letting us see how life flexes, is lost, and moves beyond. What will life look like as this age unfolds? How much biological, and cultural, evolution will take place? How will the new weather trigger the ecological reordering of natural communities everywhere as species shift/adapt/ collapse under the weight of change?

Bang, bang, bumpity bump, I hear as spruce cones fall on the metal roofing of my woodshed. The sound ricochets into the house, and my back straightens in response. The squirrels are launching spruce cones from the tops of nearby conifers, and their aim is good.

The red squirrel is a rascally creature. Its burnt-orange color and white belly are vivid identifiers, and I momentarily catch a dark, beady eye when one sits hunched on a pile of wood in my yard. Red squirrels inhabit hardwood forests interspersed with pine and spruce trees, and they feed on a variety of seeds and nuts that they collect and store for winter. I've often stumbled upon their caches, underneath the garden shed, spilling from the children's tree house, or in hollow mounds in the woodshed, where gnawed cones tumble out when I grab kindling on a wintry day.

Stan Boutin lives with red squirrels much of the year. He has spent more than a decade of his summers in their company, tracking their movements, observing their behaviors, and generally overseeing the red squirrel society in a remote region of Canada's northern boreal forest. As a professor of biology at the University of Alberta, Boutin was initially interested in animal behavior and how family relationships developed and matured over time. Boutin's research involves the tricky work of following individuals, climbing into squirrel nests and stretching his arm into the hole of a tree where the wee pups lie sightless as the mother calls alarmingly from above. Boutin weighs, counts, and takes a tissue sample from each of the young. A month later, he returns to pierce the soft tips of each ear with a colored tag so he can observe them from a distance. Working with colleagues, Boutin places hard plastic collars on the females, attaches tiny radio transmitters to their backs, and tracks the distance the squirrels disperse from their nest. In this way Stan Boutin has collected data on the behaviors of more than five thousand squirrels on eighty acres in the Yukon. He has constructed a genealogy of rodent relationships, showing that both phenotypic plasticity and genetic change are occurring as global warming settles in on the Arctic landscape.

Red squirrels are fiercely territorial. This is plain to me on a late March morning when both squirrels and chipmunks are active in my yard. Squirrels are bullies to the chipmunks, just-roused and lethargic from their quasi winter dormancy and thereafter at a clear disadvantage. The aggressive squirrels establish their territories around high-yielding spruce trees and the cones that can be harvested from them. Adults stake out the trees the way hummingbirds defend populations of red trumpet-shaped flowers or the way black bears guard berry patches, leaving their scent and trambles behind.

Squirrels mate in winter and have their pups in the spring. The pups wean in the summer, and then the offspring disperse. But these days, early arrival of spring in the northern hemisphere is quickly changing the timing of the red squirrels' life cycle. As the Yukon warms, female squirrels, Boutin found, are giving birth eighteen days earlier on average than when he began studying them twenty years ago; pups are now born beginning in March rather than in April.

Having seen that females exhibit some variability with respect to breeding date, Boutin investigated whether advancing birth time was due to the squirrels' plasticity; that is, their ability to adjust their mating and birthing to local conditions. As it turns out, spruce trees were responding positively to warmer temperatures and producing more cones, a resource boon for the squirrels. Alternatively, advanced birthing times could be an evolutionary event and, as with mizuna, genetic change might be taking place.

In conducting such a long-term study of female squirrels and their young, Boutin and colleagues had a robust data set to draw from. What they found was this: greater abundance of spruce cones contributed to the advance of breeding; greater cone supply resulted in better nutrition for females, and they birthed as many as three and a half days earlier per generation because of it. But genetic evolution was also occurring. Selection favored females who mated and birthed earlier, as their offspring were more likely to survive the winter. In just six generations, red squirrel breeding date had advanced by two weeks, a result comparable to mizuna's response to drought thousands of miles away.

Contemporary climate change, brought about by humankind's emission of carbon dioxide, is occurring along two
major axes: first, the concentration of greenhouse gases
in the atmosphere (carbon dioxide and methane being
chief among them), and second, temperature. As the
atmosphere changes, parallel changes are occurring in
precipitation, storm events, and ecological interactions
including competition, predation, and parasite/host as-
sociations. But temperature is the dominant factor of
climate change, the one humans readily feel and other
organisms are tracking over time.

Temperature influences the distribution of species
across landscapes. Sugar maples do not grow south of
Connecticut, because their seeds need freezing cold
temperatures to germinate. Apricots are rarely grown
in Vermont because the deep-winter cold freezes their
branches and late-season snowstorms kill their blossoms.
All of life is distributed across the landscape in accor-
dance with the range of temperatures organisms have
adapted to, the ones offering the best chance of survival.
There are other factors that will fluctuate as a consequence of climate change, but in many ways the rise in
temperature describes the overall phenomenon. And,
with few exceptions, temperatures are increasing across
the world.

As temperatures rise, the distribution of species
will shift. Once they arrive in new habitats, evolutionary change will follow as animals and plants adjust and
encounter novel conditions. For instance, when sugar
maples move north into what is presently Arctic tundra, they may be suited to the minimum and maximum

temperatures of that landscape but not to coexist with the lemmings gnawing at their roots. In time, perhaps more rapidly than in the past, maple trees will evolve traits that make them more fully adapted to the new conditions.

⟶∾⟵

When I ponder adaptation, my thoughts unintentionally drift to the larvae of the hawk moth, a tropical insect. Most tropical moths and butterflies are known for their iridescent colors and bird-sized wings. The morpho butterfly flashes a gorgeous spectrum of blue, violet, and aquamarine, shimmering like liquid glass and edged in a silvery brown. But it is the wingless larvae of the hawk moth that astonishes me. At first the caterpillar appears indistinct: an ordinary ivy-green body with a slender, muted-green head that extends in one direction while its similarly green tail widens in another. When a predator approaches, everything changes. Its tail expands, fills with air, and widens into the shape of a snake's head. Folds of skin mimic a snake's gaping, ominous mouth, and black spots outlined in white with yellow edges look just like a snake's pupils and menacing wide-eyed stare. There is even a slight crease between the two "eyes," imitating the indentation of a reptilian skull, an impressive feat for an animal that has no bones or cartilage at all.

The hawk moth's near-perfect mimicry elicits the kind of awe I feel when I see a young violinist play a complicated piece, except in this case it is natural selection, not talent and hard work, that has fashioned the tissue of a caterpillar to look, with exacting detail, like the predator of the very animals that eat it. But natural selection does not result in perfection, though it is hard to not see the near-faultless match of the caterpillar's tail and the snake's eye as a form of that. Rather, selection works to help organisms adapt to local conditions. As conditions change, snake predators included, corresponding changes in the organisms affected by them will change too.

My amazement with adaptation does not stop at caterpillars disguising themselves as snakes. Have you seen the orchid mantis with petal-pink legs? It so resembles the Asian orchids it lives among that only on close inspection of its "silky petals," floral pink with two cerulean dashes, do I see the mantis's eyes, blinking.

These marvels of nature and evolution do not end with insects. The outrageous and at times freakish adaptations of thousands of organisms—the head of the hammerhead shark, the luminous blue of a jay's wing, the penny-sized opossum (born hairless) clutching its mother's nipple for six weeks—are all testaments of evolution, to the infinite tinkering of biology and the constant tuning of organisms to shifting conditions.

"*Time* and *favorable conditions* are the two principal means which nature has employed in giving existence to all her productions. We know that for her time has no limit, and that consequently she always has it at her disposal," wrote Jean-Baptiste Lamarck (1744–1829). A French biologist and philosopher, Lamarck studied a range of organisms but gravitated to invertebrates: mollusks, worms, and crustaceans especially. Peculiar and unusual adaptations inspired early biologists like Lamarck and others to question the prevailing worldview of natural theology, where an omnipotent god was responsible for designing and creating each and every species.

As a predecessor of Darwin, Lamarck reasoned that adaptations were not created by a single supreme being but were instead derived from the tendency in nature to become increasingly complex, moving toward a trajectory of perfection that the maker ordained but did not establish. Life, believed Lamarck, responded to innately felt needs, what he termed *sentiments intérieurs*. In time, *adaptation* became the word to describe both the processes and the result of nature's responses to local conditions. Lamarck's thinking, initially refuted, was later supported on several accounts.

The first of Lamarck's tenets was that the diversity of life, understood in the eighteenth and nineteenth centuries from the fossil record and the travels of early naturalists to the southern hemisphere, was a result of adaptation to varied environments. The second tenet was that intricate relationships between organisms and their environments would take time. My amazement with the elaborate resemblance of the hawk moth and orchid mantis was shared by Lamarck when he saw similarly startling examples of ecological adaptation. Earth, Lamarck radically calculated, must be very old, far older than its biblical age of approximately six thousand years. Lamarck also reasoned that traits must pass from one generation to the next, what he called the "inheritance of acquired characteristics." He used the notions of use and disuse to substantiate his reasoning. A blacksmith's bicep grows as he uses his arm muscles to swing hammers and anvils. Similarly, a giraffe's neck stretches to reach the upper leaves of a tree. According to Lamarck, offspring of the blacksmith and giraffe would have larger biceps and longer necks respectively because of their use in the previous generation.

While Lamarck had his disciples, others were skeptical. There must be limits to an organism's ability to grow in a particular direction, they insisted. After all, the children of blacksmiths weren't that much different from other children, nor were the necks of young giraffes markedly different from those of the previous generation. The skeptics cited the philosopher and ecologist Goethe, who said, "Es ist dafür gesorgt, dass die Bäume nicht in den Himmel wachsen," which translates to, "It is ordained that trees cannot grow to heaven." People knew that limits exist under all sets of conditions.

While Lamarck is credited for his description of adaptation and his reasoning that the world must be very old, he was wrong about the mechanism by which a parent's traits were transmitted to its offspring; he invoked the improbable idea of "ecstatic fluids." A half century later Charles Darwin theorized that adaptation was driven by the process of natural selection. The adaptations Lamarck was so enchanted by were the result of this process. Over time, conditions changed and species diversified. While both Lamarck and Darwin wondered how organisms cope with changing environments and used the term *adaptation* to explain what they saw—the coupling of organisms with their environment, the diversification of organisms over time—their emphases differed. Now in an era of warming, where organisms experience suddenly changing environments, we see how both men were correct and how seminal adaptation is to the evolution of life.

The green bow of the canoe cuts effortlessly through the early September waters, which are clear and clean and warm enough to swim in. The sixty-acre Kettle Pond is surrounded by thirty thousand acres of state park and managed forests, enough wild country to distance oneself from the sounds and activities of modern life.

We are a party of nine, two families celebrating the end of summer. The sky is crystalline blue with barely a horsetail cloud to shield the sun. Here and there trees are turning color: a cardinal-red sugar maple, a honey-colored birch. The seasons are shifting, but this day has enough heat to land it squarely in Indian summer.

From the canoe we can see lean-tos built by the Civilian Conservation Corps in the 1930s scattered along the shores of Kettle Pond. Vertical logs painted rust-brown still support moss-covered roofs more than seventy years after they were built. The deep fair-weather structures look like wooden caves, and protect campers on three sides. The solidly built stone fireplaces have cast-iron grills that invite our day's catch. Here and there campers have grouped logs and boulders to furnish the sites so that people can eat, talk, and read overlooking the pond, *en plein air.*

We pull up at the shore and beach our boats. The children bound out of the canoe and race up to the campsites. Like real estate agents, they tout the benefits of each camp to one another: this one for its view of the pond, this one for its privy, that one for the velveteen moss-covered stump, a kind of cushioned outdoor chair. They play at cooking over the fireplaces, then lie down in the lean-tos. They can see themselves here forever. Why not? The weather is perfect; there is food in the backpacks and an extra sweater around their waists. They have all they need.

The canoe party begins to wander around the pond. A
population of pitcher plants lives at the pond's farthest
point, and several of us head toward it. Here too, in this
watery niche, rapid evolution in response to a time of
warming is taking place.

The hood of the pitcher plant is lined with stiff hairs
that point downward. It traps inquisitive ants, flies, and
moths that land on its brim by luring them with pungent
nectar. The strategy is ingenious: insects, attracted by
the scent, get caught in the plant's hairs and eventually
fall to its base, where they are digested by quick-acting
enzymes, a mix not unlike human pancreatic juice with
a trace of hydrochloric acid. As the insects decompose,
their bodies release essential nutrients and organic compounds, more nourishment than what can be gotten
from decay alone, nourishment that the pitcher plants
have ceased to gather for themselves.

Before long we come to a place where a peninsula
of sphagnum moss and drifting debris extends from the
shore. There, almost concealed by the knee-high blue-
berry bushes and lanky sedges, are a handful of pitcher
plants. Late-morning sunlight illuminates their nearly
translucent vases, bulbous bases that look marine rather
than terrestrial. The shape of the purple pitcher plant
is easily recognizable: a squat cylinder of fused leaves
streaked with red veins and finished with a flaring hood.
Similar to the unique morphology of other carnivorous
plants, it is immediately strange, yet familiar—the purple
pitcher plant is widely known and its distribution ranges
from the Gulf of Mexico to Newfoundland in places
where there are acidic wet soils.

Insects are dying inside the pitcher plants at Kettle
Pond. I can see bodies adrift in the mixture of rainwater,

nectar, plant enzyme, and the formic acid released from the trapped and decaying ants. Other insects are tangled in the hoods' hairs, where, in time, they too will exhaust themselves, then drop, drown, and decay.

Amazingly, though, other insects make a safe and prosperous life for themselves in the belly of the pitcher plant. Larval blowflies feed on the cadavers of animals that drop to them. The blowflies, equipped with claws that provide a footing in the hairy, slippery interior, lay their eggs in the pitcher's pool. The grubs hatch and develop promptly. Then they escape by boring a hole in the side of the pitcher. They pupate in the nearby moss and return to pollinate the plant as a winged adult.

Another insect happily living inside is the pitcher plant mosquito. Like the blowfly, its life plays out in a commensal relationship with the pitcher plant; the mosquito benefits from the watery habitat by feeding on decaying insects and spiders and craftily avoids the pitfalls so deadly to others.

The biology of the pitcher plant has long fascinated naturalists. Darwin's *Insectivorous Plants,* a 450-page account of carnivorous flora, was published in 1875. It was Darwin who first questioned why carnivorous plants have comparable methods for trapping insects, regardless of their taxonomy. (This idea of coinciding forms from unrelated species was eventually borne out in Darwin's theory of convergent evolution.) It is not uncommon for leaves to fuse together, creating the bulb in the case of pitcher plants and serving a variety of purposes. For instance, species of *Dipsacus,* commonly known as teasel, have leaves that merge together at the base of the stem to form a cup that collects water. This mini reservoir deters aphids and other plant-eating pests from climbing up its stalks, much like a can of water placed at the base of a post will deter termites. Leave fusing, it appears, is a multitasking adaptive strategy that serves as more than a basis for catching nutritious prey.

Throughout the spring and summer the pitcher plant mosquito completes several generations. Adults mate and lay eggs in the fetid water. Their larvae emerge, feed, and become adults. And then there is mating again. But, like the monarch butterfly whose last generation in North America forgoes mating for migration, the last generation of pitcher plant mosquitoes hatch but do not develop. Their biological clock urges winter sleep rather than growth when the daylight ebbs to fifteen hours a day. The season's last generation of mosquitoes overwinters in the freezing water at the plant's base.

Christina Holzapfel and William Bradshaw have studied pitcher plant mosquitoes in the lab and in the field for thirty years, since their graduate school days mucking about in the bogs of the Northeast. Having published dozens of scientific papers on biological clocks and pitcher plant mosquitoes, this husband-and-wife research team is now able to offer some insight into how climate change is affecting an organism that uses temperature and day length as cues to trigger its life cycle.

Here is the way William Bradshaw sees it: there are two great cosmic rhythms that organisms living on Earth's surface detect. One is the daily rhythm of Earth rotating on its axis, manifesting diurnal and nocturnal conditions over a twenty-four-hour period. The second is the annual rhythm of Earth's rotation around the sun, the rhythm that results in the longest day, the longest night, and the two equinox moments in a given year. All terrestrial multicelled organisms, as well as a handful of single-celled forms, are known to keep track of these rhythms using an internal counter that follows changing day length as an adaptive response to a consistently twenty-four-hour environment.

As organisms that follow our own biological clock, humans can intrinsically relate to this concept. We are perceptive to the daylight changing as we pass the autumnal equinox and move decidedly into the darkest part of the year. We track the declining light, listening hard for the weatherman's report of when the sun will rise and set, happy for the moment in late December when we start gaining light again. For animals, however, changing day length is not limited to just a psychological cue of seasonality; it is a reliable signal that biologically and chemically shifts them into another phase, an exact cue more dependable than temperature or humidity.

Pitcher plant mosquitoes, living in their aquatic microcosms at the base of the purple pitcher plant, are not, as yet, stressed by warmer temperatures. They are not cooling themselves or finding shaded plants to lay their eggs. There is no evidence that their physical selves are under selection because of climate change, even though temperatures are rising around them. Rather than adjust their tolerance for rising temperatures, selection is working to favor mosquitoes that remain active well after others have entered their winter sleep.

In insects, winter dormancy is called diapause. It describes the time of quiet biological activity when these cold-blooded organisms are inactive. What Holzapfel and Bradshaw have found is that pitcher plant mosquito populations living above the 46th parallel are delaying diapause; they begin their winter sleep seven and a half days later than they did in the 1970s, shifting into it when day length reaches fourteen hours rather than fifteen. And because Holzapfel and Bradshaw know their study organism so well, having sequenced the insect's genome, they also know the regions on three of the mosquito's six chromosomes that are responsible for diapause. This is where selection is acting, they argue, and the driver is climate change.

William Bradshaw is talking about the importance that pitcher plant mosquitoes get the winter transition right. "If you take the early train you use up all your resources before the season ends. The trick is to take the last train out of Dodge. If you miss it, that's it, you die."

Like the red squirrels living in the Arctic who time the birth of their pups to coincide with earlier springs, pitcher plant mosquitoes are adapting to the novel conditions by evolving a new prompt for their biological clock. Only recently, remaining active when the days contained fourteen hours of day length would result in "missing the train." Now, entering diapause at fourteen hours, rather than fifteen, capitalizes on new resources, allowing the mosquitoes to take advantage of global warming by continuing to develop and reproduce and evolve a new tolerance for shorter days.

These cases represent the first instances we know of how the natural world is genetically adapting to climate change. From these stories we can make several conclusions. First, contemporary climate change is a strong selective factor, strong enough to drive visible evolution that researchers can track. While paleontologists knew that climate changes in Earth's past have acted as selective forces, we now have evidence that the same is true for contemporary times. As long as there are genetic differences between individuals, selection can act.

Second, the traits under selection must be heritable. Third, organisms will exhibit different "strategies" depending on other sets of behaviors they also rely on. Ultimately, evolution will follow the best fit. For mosquitoes and squirrels, changing what triggers their diapause and birthing schedules has been the mechanism. For mizuna, adapting to drought by changing growth and reproduction has been its strategy. What we don't know is where the thresholds lie, how much genetic variation exists for key traits or how difficult it may be for longer-lived species with fewer offspring to respond evolutionarily to climate change. A polar bear has two offspring a year; a pitcher plant mosquito has hundreds per generation. Evolutionary change is more rapid in short-lived organisms with large populations, perhaps better allowing them to keep up with the rapid rate of environmental change that is occurring.

~~~

Extinction is an important process in evolution. It is the failure to respond adaptively to rapidly, sometimes catastrophically, changing conditions. For the species that died during mass extinction events in the past, adaptive evolution failed them; they were simply not able to respond quickly enough. How do we reconcile the potential for evolution in the Age of Warming with the fact of extinction? Will populations be able to respond rapidly enough? Are we entering an extinction vortex, as some have concluded, or is there more adaptive capacity than we realize?

Modern humans are roughly one hundred thousand years old, a relatively brief age compared to other forms of life. During that short time, however, we've evolved from small bands of hunter-gatherers who moved as resources changed in the landscape to larger populations that settled and lived agriculturally. While remnant populations of hunter-gatherers and pastoral societies still exist (e.g., the Hadzi and Mongolians), humans are increasingly urban; 50 percent of us live in a city now and by 2050, 70 percent of the world's population will be urban.

For human beings, the definition of evolutionary adaptation is different from that in the nonhuman world. For us, natural selection works *culturally and biologically* to affect behavior. The science of human genetics has shown that culture can be a powerful force for, and buffer against, selection.

Take for example how our diet—a cultural choice—has affected us genetically. Ancient Europeans began drinking cow's milk after infancy approximately six thousand years ago, when dairy farming began. This ability was conferred by mutations in the amylase protein, a metabolic feature that breaks down calcium and other milk proteins. Mutations in amylase resulted in lactose tolerance through adulthood, a tolerance for milk that normally switches off with weaning.

Anthropologists theorize that people who received extra nutrition by drinking milk had more-successful offspring. More-successful offspring with the milk-tolerant gene allowed for the gene to spread. This is the evolutionary explanation for why most descendants of northern Europeans can drink milk (and eat ice cream) as adults. Further, by examining the human genome across races (including African pastoralists), researchers found that the mutation responsible for consuming milk as an adult occurred no fewer than four separate times.

Ironically, culture can also buffer us from natural selection. Medicine, for instance, protects us from the fatal
effects of disease, while farming allows us to store food
and subsist through times of famine. Through technologies and cultural practices like these we realize that culture itself can be a selective agent.

Genetic change can be mediated by cultural forces
other than medicine, farming, and diet. Benjamin
Voight and his colleagues conducted a genome-wide scan
of groups as diverse as East Asians, Europeans, and Yo-
rubans, a people of sub-Saharan Africa. They found that
genes involved in the sense of smell, fertilization of eggs
by sperm, the metabolism of carbohydrates, salt sensitiv-
ity, brain development, skin pigmentation, and skeletal
development were all under strong selection. All are selection events that occurred during the last ten thousand
years, when humans transitioned from hunter-gatherers
to farmers. While some of these genes have been under selection for a long time, namely fertilization and
olfaction, others exhibit very recent genetic change, specifically skin pigmentation and metabolism. Voight hy-
pothesizes that these changes may reflect adaptation to
novel conditions, the transition to agriculture, and past
warming events in particular.

This leads us to ask: if culture can be a selective agent,
how can it be used to help us adapt to new environmental conditions, ones we'll see in the Age of Warming?
Like mizuna, red squirrels, and pitcher plant mosquitoes, our adaptations will shape our fit to local conditions. In the past, the choice of technologies had real
biological repercussions for the people who adopted them.
Take for instance the design of a canoe. Deborah Rogers
and Paul Ehrlich, biologists at Stanford University, report that Polynesian canoes were shaped by symbolic as

well as functional design. Their evidence suggests that Polynesians who too heavily crafted their boats with symbolism (aesthetic, social, and spiritual decorations) were less successful than individuals whose boats were crafted with functional elements (those that determine a boat's seaworthiness). The researchers quote the French philosopher Alain (Émile-Auguste Chartier), who in 1908 wrote, "Every boat is copied from another boat. . . . Let's reason as follows in the manner of Darwin. It is clear that a very badly made boat will end up at the bottom after one or two voyages and thus never be copied. . . . One could then say, with complete rigor, that it is the sea herself who fashions the boats, choosing those which function and destroying the others."

Conditions are radically changing. And while our human culture does not rely on boats alone, our success will be measured by our ability to adapt to the dynamic conditions at hand. Climate change is nothing less than the opportunity to fit in with our environment.

CHAPTER 3

On Migration

Dead Creek gently meanders in an oxbow pattern through Vermont's fertile farm country before spilling out into Lake Champlain. Along it stretch miles of land planted in feed corn, forage for the dairy cows kept in the nearby barns. Hedgerows and farmhouses punctuate the pattern of rectangular fields of grain. Occasionally a herd forages in an open pasture.

Since the turn of the twentieth century, biologists have recorded the arrival of snow geese at Dead Creek in the fall. Amateur birders flock to see clamoring geese drop from the sky. Their loud *whouk, whouk, whouk* ricochets across the valley as family groups of geese call to one another, land, and settle into feeding.

A mix of factors draws snow geese to Dead Creek and the Lake Champlain valley. Fields of corn stubble feed them, and there is plenty of open space that allows thousands of the white-winged birds to descend together. And a wide lake provides direction as southern-flowing water leads to coastal wetlands and marshes where the geese overwinter. At Dead Creek, the birds fuel up for the second leg of their fall migration that began in the Arctic and will end in a marshy patch thousands of miles to the south.

For two weeks I have heard geese overhead, honking to each other and flying in the classic V formation. Each time I hear them I stop and search the gray sky for flapping birds. I finally line up the sounds with the dark shapes of the birds' bodies as they move swiftly through the air. I count by tens, and estimate four hundred in one flock, nine hundred in another, only fifty in a third. I try to keep track of them as long as I can until they evaporate into the distant sky. Yet even when I can't see them, I can hear the lead birds reporting back to the flock.

Hearing geese migrate is a signal. Like a deciduous forest changing colors or the way onions sweeten after the first frost, it announces that autumn has arrived. The age-old pattern of leaving an area when local resources grow scarce has begun. This place is no longer suitable; it is time to move on.

The day is clear and cloudless when I arrive at Dead Creek in mid-October. It is close to peak migration and not uncommon to see twenty thousand snow geese in a single field. A dozen cars are parked at the refuge kiosk, and I see several people with binoculars around their necks. Others have set up cameras and tripods, hoping to capture the expanse of white birds milling and pecking. I walk to the kiosk but already know from the absence of sound that there is nothing here. There are no geese, only observers looking up into the soundless sky. The birds are late and we are asking why. Perhaps warmer conditions to the north have kept them there. Or maybe agriculture has expanded in upstate New York and southern Canada and there is more corn stubble in the fields persuading them to remain longer. It could be a combination of these factors, or neither. We do not know.

Migration is a long-established strategy in human and nonhuman worlds. Songbirds migrate to overwintering grounds when food grows scarce in the north. Monarch butterflies migrate south to hang torpid from the boughs of tropical evergreens when northern milkweed leaves brown and die back. African pastoralists, nomadic Tibetans, and Minnesotan "snowbirds" migrate annually too. All are seeking better conditions: weather, food, forage, water.

Human and nonhuman migrations are linked as adaptive strategies in response to changing conditions. As Earth warms, new migration patterns and alterations in old ones are occurring. Snow geese arrive later to Vermont's cornfields, and songbirds—wrens, warblers, and thrushes—shorten their flights south. Fish and insects are migrating differently too, seeking climate spaces that allow them to mate, flower, or otherwise behave as they do now. Organisms are adapting their migrations to a northern climate that is more suitable during winter, a season that is no longer the death sentence it once was. Some, like snow geese, are staying later in their northern breeding grounds. Others, like black brant geese, are forgoing migration altogether.

Environmental conditions are not static. They vary and organisms adjust. Over the last twelve thousand years, since the advent of the interglacial period known as the Holocene—the time when agriculture arose in human societies and large Pleistocene animals like the woolly mammoth and saber-toothed tiger went extinct—countless migrations have occurred. As glaciers retreated, animals and plants ventured north to colonize newly ice-free lands. Many species expanded their ranges once the ground and waterways were let loose from the ice. Others remained in the south most of the year but ventured north to take advantage of the flush spring and bountiful summer. Thus the evolution of seasonal migrations in animals arose as a way to increase the number and fitness of a species' offspring; southern species could increase their progeny by exploiting the fresh growth and abundant prey in northern latitudes and then return to more benign conditions for the remainder of the year.

Migration in North America is often thought of as the movement of animals from the south to the north and back again. But there are also shorter migrations that happen within the North American continent itself. Carolina wrens, for instance, migrate south, but only to Maine and Vermont after spending the summer in Canadian forests, and Arctic caribou migrate above the Arctic Circle to feed on the new young growth of tundra shrubs and birth their young before they return south to winter in Canada's northern forests.

As the planet enters the Age of Warming, migratory animals are accommodating changing conditions, the very evolutionary stimulus from which their migratory behavior arose. Biologists are observing animals as they alter their historic migrations to avoid trophic mismatch, which is being in the once-correct location but at a time when expected resources have passed. Caribou, which have timed their migration to correspond with new plant growth, are finding that the earliness of spring prevents them from feeding on the most nutritious young leaves, the value of which lactating females pass on to their young.

Similarly, the great tit, a songbird that migrates from North Africa to the United Kingdom each summer, times its migration to feed on the larvae of the winter moth. Historically, great tits hatched in synchrony with the abundance of their preferred prey. But with spring's advance, the moth, which times its emergence with young oak leaves, has also advanced and is laying its eggs earlier. Its larvae are grown and gone by the time the great tit touches down from North Africa, leaving it with an empty plate. Like the caribou's arctic shrubs, the winter moth is responding to early spring conditions and upsetting the table before the great tit has even arrived. Species involved in these types of trophic mismatches, where migration has evolved to be synchronous with resources, are the ones at greatest risk. Paradoxically, they are also the ones under the greatest selective pressure to change.

In December 2008 brown pelicans, migrating south from the Northwest Coast to Mexico's Baja Peninsula were caught in a winter storm. Hundreds of pelicans fell from the sky. Wildlife rehabilitators combed the beaches and picked up the gangly birds as they drifted in California's waters, disoriented and dying of frostbite and starvation. The pelicans, which typically migrate out of the Northwest earlier in the fall, had remained to eat the booming numbers of anchovies and sardines, salty fishes they consumed with gusto. A windfall that became their downfall.

Pelicans have been on the planet for more than thirty million years, and their impossibly graceful bodies are a visual testament to their long evolution. The pelican evolved from the archaeopteryx, a crow-sized dinosaur with a lizard's tail and joints lined with feathers that is thought to be the link between reptiles and birds. Known only from fossils (and initially only from fossilized feathers), the flying dinosaur was discovered in a German quarry in 1860 and later described by the English biologist Richard Owen. Charles Darwin, having published *On the Origin of Species* in 1859, used archaeopteryx to explain "transitional forms," species that existed between contemporary times and their primitive ancestors.

Seeing pelicans today, diving as they do to dip their pouch like a fishnet, gulping back tens of fishes with one elegant catch, helps us imagine a time when Earth was altogether warmer and animals, in general, were altogether bigger. When pelicans fly, their seven-foot wingspan glides but mere inches above the ocean surface, an image that evokes birds of mythical lore. Like horseshoe crabs, pelicans have evolved curiously little since they first originated. In a world where extinction characterizes life's diversity as much as adaptation, pelicans represent those left standing, the ones that have not succumbed nor become greatly differentiated in response to the vagaries of changing climates. Perhaps the Age of Warming will change all that.

<hr />

Baja's Pacific coastline is quieter than the rocky landscape along the Sea of Cortez on its eastern shore. Salt flats extend for kilometers when the tide is low, reflecting the near-constant shimmering sun. Shorebirds by the tens of thousands flock to these salty flats, rich in aquatic plants and marine invertebrates, to feed on eelgrass, crabs, shrimp, and mussels. They drop from the sky, legs extending from an aerodynamic tuck near the belly. Wings flutter, ready to fold back into standing or walking pose. I look out over the mixed flock and see black-bellied plovers, curlews with strangely curved bills, and quickly moving sanderlings, petite birds that chase the retreat of waves, pecking at sand fleas and brine shrimp left behind on the wet sand.

As the shorebirds move south, lumbering among them are geese. Pacific black brant are also making their way to Mexico to feed on eelgrass, *Zostera marina*. Brant are dark, stout geese with black heads and matching black necks decorated with a delicate fan of white feathers. Their bellies are white with tawny brown edges, and like other geese they are gregarious and fly low in ragged flocks.

When black brant hatch in the north, biologists tag some with hard plastic bracelets on their legs and necks. Engraved with a unique set of colors, numbers, and letters, the tags can be read from a mile's distance with a powerful telescope. The information imparts where the bird was born, who its family members are, and a host of statistics specific to the individual bird gathered by researchers (and hunters) as they sight the bird throughout its range. As with Boutin's arctic squirrel, collecting long-term data on a species informs us about how animals' behavior is changing as the planet warms. For a migratory goose, biologists ask: will brant still migrate when their summering grounds remain habitable in winter? Like the pelicans, might they get caught in warm and then suddenly cold air?

David Ward, a bearded man with soft brown eyes and a slight build, is most at home in the wetlands of North America. He is a biologist with the US Geological Survey in Anchorage, Alaska, and has been studying Pacific black brant for decades. He has banded them in their nesting grounds in the Arctic, and then followed them south to the largest eelgrass bed in the world, the Izembek National Wildlife Refuge in Cold Bay, Alaska. Here brants gather in the tens of thousands and prepare to fly south. From Cold Bay, Ward follows them down the Pacific coast to sheltered bays along the Baja shoreline until the geese reach their southern terminus in Sinaloa on the Sea of Cortez.

Ward knows brant geese the way others know their household pets. He can distinguish between calls that signal predators and those that alert the flock to "lush eelgrass ahead." In 2009 Ward published a paper titled "Change in Abundance of Pacific Black Brant Wintering in Alaska: Evidence of a Climate Warming Effect?" in which he showed, surprisingly, that for the first time in his forty-year study of brant populations, some are no longer migrating. More and more geese were staying north, overwintering in the shallow subarctic bays where eelgrass now grows even through the harsh conditions of an Alaskan winter.

Brant geese living along the Pacific coast (there are Atlantic brant that migrate along the eastern seaboard) follow what David Ward calls "the green wave" as eelgrass beds become productive and the birds seek them out. For years the geese would breed in the high Arctic and then begin a southward migration, fueling up on eelgrass at the Izembek refuge before they took their nonstop flight across the Gulf of Alaska, a two-thousand-kilometer trip. But with the Arctic warming and the number of days that ice covers the bays falling, the eelgrass remains edible later into the year. The brant began "reconsidering" their migratory impulse. Populations made up of family groups—parent geese traveling with their juvenile offspring—began "short stopping," gambling that they could reserve their resources and withstand an Alaskan winter rather than make the energetically taxing southern trek. In essence, the geese were displaying a plastic response by greatly shortening a migration most researchers consider thousands of years old.

There are two factors that help explain why the black brant come together in the hundreds of thousands at Izembek. First, Izembek contains the largest eelgrass bed in the world, and the eelgrass there is more nutritious than in Mexico. Second, the Aleutian low, a low-pressure system that brings southern winds to the region, has historically sent brant scurrying out of the cold Arctic. There's nothing like a strong tailwind to launch a transoceanic flight! Typically, the Aleutian low is centered at 55 degrees north latitude, the same latitude as Izembek. But as the Arctic warms, the Aleutian low shifts northward into the Bering Sea. Now it lines up with brant migration only sporadically; it is no longer a dependable clue for a bird sorting through environmentally crossed messages like too many lights blinking on a switchboard.

At its least complicated, adapting to climate change means finding ways to adjust in the face of changing local conditions. Migration, and its inverse, not migrating, can be the agent of adjustment. Geese that no longer migrate are responding to easier winters. They are overwintering in regions closer to their breeding grounds rather than expending energy to fly. While not migrating carries risks—ice can freeze the bays where eelgrass grows—evolutionary theory predicts that selection will favor birds whose behaviors best fit their circumstances. If remaining in Alaska is the fittest strategy, selection will favor the ones who choose it. Alternatively, if migrating only part of the way to Baja conserves energy and hedges against variable winter conditions, we may see brant wintering in, say, Northern California. Here too natural selection will act.

Salmon migrate from the ocean to the headwaters of rivers and streams, where they breed and spawn. Freshwater eels do the reverse: they journey from inland waters to breeding grounds in the Sargasso Sea, a region in the middle of the North Atlantic Ocean. Geese, salmon, and eels are among the thousands of species whose migration patterns crisscross Earth each year, like Manhattan commuters on the maze of subways, attempting to be in the right place at the right time. Now, shifts in terrestrial, aquatic, and oceanic landscapes cannot be pegged to changes in light or length of day. The differences between the seasons are blurring, and while many species still respond to day length, others are finding that the waters are still warm, the leaves still on the trees, and their prey are still swimming. Why go?

The question of whether to migrate is not for animals alone. In 2007 twenty million environmental émigrés fled unsuitable environmental conditions in their homelands, outnumbering, for the first time, refugees fleeing from war. Now differentiated from other types of migrants, environmental refugees are leaving their homes because of floods, hurricanes, and droughts strongly associated with climate change. Soon, people will be migrating due to rising sea levels too, as island nations like the Maldives become submerged and their citizens are forced to migrate. By 2050, it is projected that fifty million to two hundred million climate émigrés will exist worldwide.

People have fled from natural disasters forever. Yet in the preindustrial past they responded much as they do today, by moving away from the threat and toward a place of plentiful resources where they feel less vulnerable. Then, packing up and moving on was a simpler affair. Now, in the industrial age, human populations are by and large aggregated in cities and towns. Our investment in our home places has increased as we built houses, businesses, and elaborate modes of transportation and trade. Investment in our social and physical infrastructure, and the material culture to which it belongs, has seeded a less flexible attitude to leaving a particular place even if our vulnerability grows. After the catastrophes pass, we insist on building again. The inherent response to migrate when conditions are unsuitable is replaced with the hubris of problem solving. Now we innovate our way out of difficulty: we raise the levees, we build houses to withstand tornado winds, and we engineer causeways as the waters rise around us. In many ways we find this easier than moving.

Sixty percent of the world's population lives along a coast. The people in Bangladesh's deltas, on Florida's developed coastlines, and on islands everywhere are at tremendous risk as sea levels rise. For instance, the Intergovernmental Panel on Climate Change predicts up to half a meter (18–59 cm) average sea-level rise by 2095, but others have found these projections conservative. The revised models that account for more rapid input from glaciers in Greenland and Antarctica blow away the IPCC estimates and calculate up to a two-meter rise. To put these figures in a relevant context, when Hurricane Gustav hit New Orleans in 2008, the accompanying storm surge came within one-third of a meter of the top of the levees. By the end of the century it is very possible that similarly strong storms hitting the Gulf of Louisiana would crest the levees, given the one to two meter higher seas.

Rising and falling water has historically been a catalyst for human migration. For eastern Europeans living near the Caspian Sea, fluctuating water has dictated their ability to use alternating wet and dry lands; people migrate when the waters inundate them and return when the water levels fall. Beginning in the 1930s, water in the Caspian Sea, the largest inland body of water in the world, fell by three meters. This created large new areas of coastal lands, especially in Russia's rich Volga delta.

"The change of the Volga River discharge depends principally on precipitation in the catchment area, which is controlled by climate change," writes C. X. Li, a researcher who has studied the hydrology and dynamics of the Caspian Sea. During hot, dry periods in Europe and Asia, there is increased evaporation, and water flowing into the sea from the Volga River decreases. Alternately, increased water runoff and decreased evaporation makes waters rise, but can also bring on catastrophic flooding. When waters began to rise in 1978, including a thirty-meter rise in a single year, twenty thousand square kilometers were inundated. Villages along the Caspian were drowned, and mosques, churches, bazaars, and Soviet-era apartment buildings were buried at sea.

And the people living there? Thousands resettled inland. Others migrated to America and Europe. "Migration tends to perpetuate itself, what is known as 'chain migration.' The first migrants serve as bridges between the original populations and their eventual destination," write the anthropologists Sabine Perch-Nielsen, Michèle Bättig, and Dieter Imboden. Families follow families, neighbors follow neighbors. In the Age of Warming, whole populations will migrate, seeking asylum from the rising tide, river, or, ironically enough, lack of fresh water.

Demographers predict that people will move inland or within their home countries first. Then, as land grows scarce and as populations become too dense, people will look to permanently resettle elsewhere, distant from their homelands. Cities like Burlington, Vermont, have already agreed to take them in.

Once upon a time on Earth, human migrations networked the planet the way songbird populations do today. Settlements varied with the seasons and what could be planted. People traveled in small bands in search of what they needed, and moved on again when the harvest ended or another harvest began somewhere else. Then, as agriculture took hold, fewer migrations took place. People settled, crops grew, and children were weaned on cultivated grains, allowing their mothers to have more children spaced more closely together.

Now, migration numbers are on the rise. People are searching for landscapes where resources are abundant, or at the very least, where land is available. There is a momentum behind these migrations but it is more complicated than in the past. Now there are political boundaries, quotas, and immigration offices to contend with. And yet the push is building, as more people are moving out of coastal zones, upslope, and away from floodplains. One estimate is that one hundred million people will be affected by sea-level rise of a single meter, a scenario we are likely to reach by midcentury. Like organisms everywhere, human migrants are looking for space, an unoccupied niche, a place where there is still room.

～≈～

In Great Britain, butterflies are a kind of national wildlife. Unlike in America, where coyote, bear, elk, deer, and moose roam across many portions of our landscapes, the United Kingdom has lost most of its charismatic megafauna; the red deer is the largest species. Deforestation, industrial development, and the encroachment of agriculture into wild lands have minimized the opportunity for coexistence. Recognizing this fact, British conservationists in the 1960s began putting aside land for butterflies. Simultaneously, they monitored them, studying how abundant they were, how diverse, and where their ranges extended to. British ecologists now have an expansive knowledge of their butterfly fauna, enough to feel comfortable transporting butterflies to new habitats as they anticipate climate change, practicing what has been called "assisted migration."

In July 1999 ecologist Stephen Willis collected and then released the marbled white butterfly, approximately sixty-five kilometers from its northernmost population in the United Kingdom. A relatively common species, the marbled white is the dalmatian of butterflies: dirty-white background marked with inky black spots.

Concerned that the marbled white would fail to track the changing climate by migrating upward in latitude (it is a very weak flier), Willis modeled which locales on the British landscape the butterfly would do well in as the climate warms. As the marbled white is a grass-feeding butterfly, he was less concerned about available plants for the larvae to feed on and more concerned that the butterfly couldn't move fast enough.

Human-assisted migration is not new. Wildlife biologists return animals that have strayed outside the boundaries of national parks. American bison are notorious for wandering out of Yellowstone National Park and becoming fair game for hunters. Conservation biologists have maintained captive-breeding programs for eagles, condors, black-footed ferrets, and others, and release them back into the wild when habitat conditions improve or the populations prove robust. But relocating species in anticipation of climate change is a new concept, and moving butterflies is just the beginning.

We know more about the number of galaxies in the universe than the number of species on Earth. There may be up to one hundred million or more fellow species, though an estimate of thirty million is a good and modest number. Taxonomically we have identified only a small percent of this total, but the numbers themselves reveal the magnitude of what is undescribed and unknown. Further, they reveal what it would entail to conduct assisted migration for even the smallest subset. Granted, assisted migration could aid species with low dispersal ability. It might even be necessary for species for whom migrating to new suitable habitat is impossible given the degree of uninhabitable landscape between their present and future ranges. Still, becoming agents of migration raises ethical questions and reveals how problematic it will be to conserve species in a time of warming. While assisted migration may seem like a pragmatic approach, it uncovers how little prepared we are for the wholesale disruption we are likely to see.

In 2006 the marbled white was thriving in the new colony site that Willis founded. The population had expanded during the six years since establishment, but 95 percent of the butterflies remained close (within five hundred meters) to the release site. This revealed just how immobile the species is and how unlikely it would be to track rapid climate change on its own.

The migration of species in a warming climate is upon us. Undoubtedly we will see an array of permutations as the phenomenon of climate change unfolds. Some species will extend their migrations; others will migrate to novel locations. Many will become full-time residents of areas where they used to live only part of the year.

Perhaps even we will find ourselves migrating, leaving the home places that used to be perfectly suitable but over time made us feel less secure. The human community will assist one another then as we do when earthquakes hit or tsunamis land. "We are wired," Jeremy Rifkin says, "to be empathetic. To empathize is to civilize."

Perhaps our empathy will extend to the nonhuman life made vulnerable by climate change. This will be a unique extension, given that they have been forced by our actions to need the assistance. As empathetic beings coming to the aid of species unable to adapt to the Age of Warming, we will revise our role in the ecological world from agents of relentless environmental degradation to agents in service to life or, to paraphrase Janine Benyus, the founder of the Biomimicry Institute, as agents who create the conditions conducive to life. Perhaps in this Byzantine fashion, humans will become a new kind of mutualist, not only benefiting from, but providing benefit to, the life that surrounds them.

CHAPTER 4

Feast or Famine

Three ducks are tipped up in Ben Falk's pond. Two females, mottled blue with black speckling, feed vigorously on aquatic plants while a single drake bobs among them. His feathers are a fusion of stark white and yellow-brown, like the coloring of an Appaloosa horse. These are Ben's ducks.

Ten years ago Ben was a student of mine in environmental studies at the University of Vermont. He was in his final year and thinking of how to apply his degree in ecology to the world of agriculture and the design of buildings. In his thesis he focused on ecological design: straw bale buildings with adobe-covered walls a foot thick; green roofs that attracted nesting songbirds; and living machines, a type of wastewater treatment where microecosystems of snails, hyacinth, and worms decompose household gray water for gardens and flush toilets.

By graduation Ben was intent on homesteading. While his classmates moved on to the Peace Corps and jobs distant from Vermont, Ben stayed and honed his

skills as a carpenter, landscape designer, and permacul-
turist. We meet occasionally, and he shows me his latest
project: a portable sauna; a Volkswagen run on biodiesel
from waste cooking oil; and hardwood logs inoculated
with mushroom spore from the Northwest that sprout
shitake, pearl, and oyster mushrooms. Ben is always innovating, projecting what the future landscape will look
like and preparing for what people will need most.

"Climate is a roller coaster of changing conditions,"
he tells me as we survey his current projects. "But the
earth's system has always been one of dynamic change,
before and after the arrival of human beings. To survive
and thrive in this biosphere we have to be, and have had
to be, highly adaptive."

Ben bought a hillside property not far from where
I live. It came with a lackluster house and a history of
abuse. The land was denuded in the 1800s when Northeast forests were cut for farms and pasture. Once
cleared, it was heavily grazed by sheep, which resulted
in erosion of the land's best soils. When sheep farming moved west and hilltop farms were abandoned for
richer soils of the Midwest, white pine took advantage
of the absence of cultivators and colonized the property.
In the late 1970s the property was cleared again, this
time for a view.

In Ben Falk's world, producing food in the Age of
Warming means experimenting with "whole-systems
design." His approach to gardening and farming is to
grow plants in dense clusters, cultivating diversity at every turn, choosing varieties for their capacity to do well
in extremes: hot and cold, dry and flooded. Ben works
on the faith that humans can be instruments of restoration. "We can generate ecosystem health and grow food

simultaneously," he asserts, an appealing ideal that he is tending on a mountain in Vermont.

Foremost among Ben's principles is water and soil conservation. Depleted soils, like the ones on his property, are restored with compost and by planting green manures (nitrogen-fixing plants that effectively move nitrogen from the atmosphere to the soil). These techniques promote proliferation of life in the soil: millions of bacteria; thousands of invertebrates, such as primitive collembola and predatory millipedes; and great stretches of threadlike white fungal hyphae, mushroom filaments that can run hundreds of miles from their sources. Rich soils let water seep in, slowly coursing through the earth, where roots radiate out at all angles to acquire it.

"Water management begets soil management. This is the foundational piece that drives the movement of the other pieces," Ben remarks. We are sitting in his artfully designed timber-framed barn. Outside, a series of ponds and contoured swales capture water and release it to stands of fruit trees and berry bushes, mimicking the way narrow tributaries depart from a stream and carry water to a forest glade.

Ben's land was once a hillside of poor soils and white pine monoculture. Now it's a mosaic of planted ecosystems, perennial gardens, and beds of annual crops. His pragmatism imbues his design with beauty and practicality, in the same way the swing of a scythe naturally aligns with the motion of the hay maker's body. Like other pragmatists, Ben interacts with the world experimentally, constantly testing ideas against action. His gardens are a working instrument of his belief in adaptation.

A broad definition of *adaptation* in its social context is
the adjustment of human practices, processes, and capital in response to an actual or anticipated threat. In the
Age of Warming, social adaptation means generating
resilience when weather extremes, seasonal variability,
prolonged heat waves, and changing rainfall become the
norm. Growers like Ben, reckoning with this forecast,
remain optimistic that the way we grow food can be
adapted, to small agriculture in particular. By diversifying crops, managing the pulses of water during drought
and deluge, and using a mix of old and new technologies,
we can prepare for the changes ahead. These new farmers are schooled in the science of ecology and practice a
kind of natural-systems agriculture that mimics the plasticity and resilience found in nature.

Agriculture has always been vulnerable to weather.
The Anasazi people of the American Southwest van-
ished, in part, due to a series of droughts. Mass emigra-
tion of Oklahoma farmers during the Dust Bowl of the
1930s is attributed, in part, to a string of very dry years.
History tells us that rapid changes in local climates have
contributed to the demise of societies, especially if those
societies were already compromised by overpopula-
tion, overexploitation of resources, and social conflict.
Droughts in particular are often implicated. "The land
was in the wind and everyone was dying of hunger on
this sandbank of hell" are words inscribed on an ancient
Egyptian tablet, but they are sentiments that resonate
with contemporary communities in southwest Australia
and the Sahara Desert, where drought issues misfortune
today.

Distinct from regional aridity or periodic water scarcity, *drought* is defined as a clear departure from the

historical record of precipitation. It is a relative concept and by definition region-specific. But droughts are not only meteorological phenomena. Their full effect is felt socially, economically, and culturally. Furthermore, droughts can be hastened or lessened by the social responses to them. For instance, some communities respond to drought by building reservoirs to parse out water when supplies are low. Ironically, this strategy can draw people to the community, increasing demand and concentrating and hastening drawdown.

As the planet warms and oceans heat up, the hydrologic cycle—the movement of water from the oceans, streams, rivers, and lakes up to the atmosphere and back to the land's surface—will accelerate. Evaporation, condensation, and precipitation will intensify. The same warmth that affects oceans will dry out deserts, plateaus, and recently deforested terrain. Global circulation models predict that one-hundred-year drought cycles may accelerate to ten- to fifty-year intervals.

Sea-level rise and melting glaciers further compound the dilemma of water for agriculture. Seas are expected to inundate fertile deltas, intrude into coastal aquifers, and further limit availability of freshwater for agriculture across the world. The people of Bangladesh, the Netherlands, and Louisiana can testify that this is true. Moreover, glaciers recharge underground aquifers, flow into aboveground rivers, and feed human-made irrigation channels that farmers depend on. But as Earth warms, glaciers are disappearing and no longer supply these vital waterways as they once did. Asia's Indus River, for instance, is fed by meltwater that comes from Himalayan glaciers, ones that are projected to disappear by 2070. Similarly, glacial meltwater in the Ganges

River constitutes up to 70 percent of the sacred river's spring and fall flows.

Asia is not the only region where the uncertainty of water for agriculture is a concern. Changing precipitation in the eastern United States now corroborates climate predictions. Greater deluges and longer periods between rainstorms are no longer anomalous events. In the past five years, several historical records have been broken within the region: the wettest October, the wettest May, the warmest March, the greatest rainfall in twenty-four hours, and the largest snowstorms in recorded history are among them.

~~~

The term *permaculture* was coined by Australians Bill Mollison and David Holmgren. It combines permanent human culture with permanent agriculture, each supporting the other in an interaction that, if carefully tended, will sustain us long into the future. Permaculture is based on an ecological approach to agriculture, planting natural communities of crops where species interact mutualistically (each benefiting the other) or commensally (one benefiting, with the other seeing no effect). Rather than planting straight rows of single species, permaculturists take advantage of every dimension of planting space.

Ben's home place, called Dean's Mountain after the foothill that rises behind his house, is a permaculture site. His orchard is seeded with nitrogen fixers like clover and vetch, and they supply the trees with available nitrogen as they grow. Comfrey grows underneath the trees too, because its deep roots have the ability to pull calcium and magnesium from the soil like a kind of mineral pump. Then, when the leaves fall or are cut back, they are left in place to decompose and release minerals to the fruit trees' roots. Fast-growing poplar, planted as a windbreak to shield the fruit trees, also provides firewood and, when mulched, becomes the medium for edible mushrooms, ones that will fruit well before the orchard itself.

Now imagine chickens in the orchard, pecking about in the understory and eating the curculio beetles that damage the fruit trees. Cows graze, too, preferring the nutrient-rich hazelnuts that fall to the ground to the sugary apples. In essence, permaculture is a kind of applied ecological science and works on the principle that gardeners can connect the adaptations of their varieties to create balanced and productive living systems.

For years Ben talked about adapting food production to increasing temperatures and more variable weather, focusing particularly on how pulses of water would need to be channeled. The design of his home place emphasizes the capture and storage of rainwater, a preoccupation that has brought him to his latest project: rice paddies.

This explains the series of ponds I see at Ben's. There are five of them terraced on two acres of land, evoking a Japanese landscape where paddies with curving perimeters and irrigation channels form a series of steps down a steep hillside. Flocks of ducks roam around the ponds and among the plants, eating bugs and slugs and fertilizing the soil. The ducks like the mulched rows of thorny Siberian seaberry, a fast-growing bush that bears yellow-orange berries rich in vitamin C and is a staple plant to the permaculturists because it grows from Siberia to Connecticut. Seaberry is interspersed with currant, gooseberry, and honeyberry among the fruit and nut crops—pear, cherry, plum, hazelnut, and walnut. Annual crops are planted in pocket gardens among it all, in which the kale, chard, and daikon radish were still flourishing on the late fall day I visited.

We walk down a slope to two newly established rice paddies. These were connected to a large upper pond by a narrow spillway so that when the pond fills, the overflow floods the paddies. Bermed on all sides, the rice paddies keep the water in, much like the ancient terraces built by the Inca and the Chinese. The water drains through several feet of dark, rich soil until it reaches clay at the bottom. There it slowly percolates downhill from the paddies, watering orchards as it descends.

Ben is raising short-grain rice, *Oryza sativa japonica,* in five-gallon buckets while he prepares his paddies. His well-tended nursery plants have thrived, and he's collected thousands of seeds for next year's planting. "In these two five-hundred-square-foot paddies I'll cultivate one hundred pounds of rice, plenty for a year," he says, beaming. Other gardeners and farmers in the Northeast are planting rice too. They are enticed to bring this grain into their vegetable gardens and to experiment with cultivating a tropical plant close to home.

Linda and Takeshi Akaogi are rice farmers in Vermont. As their region warms and gets wetter overall, their conviction that rice will do well increases. Rice originated in the tropics, but humans have adapted it to grow in the temperate regions of Italy, Russia, Chile, Korea, and

the western United States. While there are only two cultivated rice subspecies in the world, *Oryza sativa indica* and
O. sativa japonica, astonishingly there are 120,000 varieties that have been derived from those two. These
have been developed over thousands of years as farmers and horticulturists selected traits that best suited
the climates of their gardens and the taste buds of their
customers. How different sweet brown mochi is from jas-
mine, sticky basmati from Italian arborio.

Takeshi Akaogi is Japanese, and his tradition with
rice runs deep. Like other Japanese he celebrates rice as
a household staple. In 2005 he and Linda began cultivating dozens of varieties of rice at their vegetable farm near
the Massachusetts border and established new protocols
for growing rice in the northeast United States. By 2008
the Akaogis boasted a yield of six thousand pounds per
acre, twice the yield for wheat.

Like Ben, the Akaogis built paddies on their farm,
spread chicken manure to fertilize the soil, and constructed shallow ponds to warm the water before it
flooded the tender seedlings. As news of their success
grew, they held workshops and taught others. With a
grant from the Department of Agriculture they grew
thirty rice varieties in their own garden before con-
cluding that the variety Hayayuki grows best in the
Northeast.

Hayayuki means "early snow" in Japanese. It was
first developed in Hokkaido, an agricultural area at
43 degrees north latitude, comparable to Vermont's 42
degrees north. After cultivating it, the Akaogis grew
Hayayuki in order to give the seed to people like Ben,
Johnny Appleseed fashion. The early snow variety now
grows in hundreds of gardens, with reports of its merits

spreading from planter to planter like rumors on a telephone party line.

In April, Ben will plant his Hayayuki seed in two-inch plugs of soil in an unheated greenhouse. He carefully covers the bright yellow-green shoots with cotton muslin when temperatures dip below 30 degrees Fahrenheit (-1 degree Celsius). Once growth has begun, Ben will flood the seedlings with warm water, keeping them moist and further moderating any late-season cold. By mid-May he'll plant the shoots in his two paddies, and by mid-July the rice will be flowering, attracting sandpipers and killdeer, thick-bodied dragonflies, and needle-thin damselflies. Spying the paddies, Canada geese will land with a muffled splash, raising a ruckus among the resident ducks. In September, Ben will cut the rice with a scythe, then dry it in a threshing barn before shaking out several hundred pounds onto a drop cloth. The brown-grained rice will be poured into bags and stored for the year's consumption. In time the frequent downpours that Ben anticipates will gainfully flood his water-loving rice rather than drown fields of less well-adapted crops like corn, squash, and tomatoes.

Aldo Leopold, the author, biologist, and ethicist, once wrote, "A thing is right when it tends to preserve the integrity, stability and beauty of the biotic community. It is wrong when it tends otherwise." Leopold's credo is frequently invoked by conservationists to encourage the protection of biodiversity. Decades of ecological research support the claim that diverse ecosystems are more resilient to natural disasters like cyclones, windstorms, and wildfires; diverse rainforests recover more quickly after hurricanes blow through, and diverse coral reefs withstand storm surges or influx from stream erosion better than reefs with less diversity. This scientific tenet applies to agriculture as well, especially during an epoch of change.

For some farmers, adapting to climate change will mean revising and/or expanding the varieties they cultivate. Corn farmers will grow rice, apple growers will grow peaches and apricots, and figs and kiwis will find a home in northern backyard gardens. Growers will track climate change as local weather reveals the broad-scale phenomenon of global warming. Some will buy seed with traits that increase their success in a warming world. Heat-shock resistance, drought tolerance, earlier flowering, and novel defenses against new pests will rank high. But where will these traits be found?

Seeking to keep humanity fed in a time of warming, botanists are turning to the living ancestors of the three hundred crops that humanity relies on. Natural populations of wheat in the Karacadag Mountains in Turkey's Fertile Crescent, thousands of varieties of maize found throughout Mexico, and native rice stocks in Thailand are being sampled for their genetic diversity. Wild-sourced plants are being hybridized with conventional ones to bring about new, more resilient varieties. In the hands of plant biologists and breeders, the ancient strains become the raw material of evolution. Just as climate change has been a natural selective force on mizuna, plant breeders select traits that confer success as conditions become more variable and extreme.

There will be a greater need for the technologies that make better use of the nearly 70 percent of the world's freshwater that is currently routed to agriculture. New irrigation equipment will likely stem evaporation losses. Desalinization plants, perhaps running on renewable energy, will burgeon as rivers that once carried freshwater to agricultural fields dry up. But beyond water technologies, there are bioengineering firms that aim to insert "climate resilience" into the germplasm of crop seeds themselves.

Monsanto, the world's largest seed company, has filed hundreds of patents on gene sequences it terms "climate ready." Like its crop products that confer resistance to herbicides (Monsanto's Roundup Ready is expressed in the crop plants themselves, conferring resistance to herbicide when it is sprayed to kill crop weeds) and to pests (through the crop plant's expression of a toxic bacteria, *Bacillus thuringiensis*), climate-ready genes in corn, soy, wheat, and sorghum will confer the ability to withstand drought. "Drought-tolerance genes would allow crops to deliver strong yields with less water," states Monsanto's website. "Drought-tolerance genes are expected to protect and possibly increase crop yields in areas with moderate drought stress, while reducing irrigation costs."

"Mega-agribusiness" technologies will not be a panacea. In fact they could be very dangerous and truly counterproductive, by removing the insurance that genetic diversity provides. While twenty-first-century technologies will play a role in how agriculture adapts to the Age of Warming, traditional farming practices may be more lasting. Certain time-honored techniques can improve a farm's resilience, including land fallowing, trenching, small-scale water interventions (like constructing paddies on a Vermont hillside or terracing crops to optimize rainfall), altering the time that a crop is planted, and diversifying on-farm activities (increasing the types of crops and colocating animal and vegetable cultivation).

Already, crops are reaching climate thresholds with the conventional genes they carry. For instance, corn and soybeans are two of the four largest sources of calories in the world, of which the United States produces 41 and 38 percent respectively. In a study that examined temperature thresholds for corn, researchers found that at 84 degrees Fahrenheit (29 degrees Celsius), corn yields were significantly reduced. Similar trends were found for soy and cotton. This suggests that critical crops for the world's food supply have a limited adaptation to climate change. There are breakpoints beyond which crops are no longer resilient.

Over the next several decades, temperate areas of the globe will experience an additional 3–5 degrees Fahrenheit (1–2 degrees Celsius) increase, and tropical areas will increase between 4–7 degrees Fahrenheit (1.5–3 degrees Celsius). "The coolest growing season of the future will be hotter than the warmest seasons of the past," writes anthropologist Cary Fowler. Yet she insists that crop diversity can buffer human societies from extreme weather. "Ironically, climate change threatens the diversity that can be used to help us adapt to climate change [itself]." Not only do we need to expand the three hundred species of food that humanity relies on, but we need to collect seed from places that have experienced the conditions the planet will see in the future, in hope of preserving their genetic material before they vanish. And not only seeds. Roots, legumes, tubers, and other vegetable matter must also be preserved.

And so people are collecting. As with assisted migration, people are motivated to protect the world's biodiversity in an era of warming. This is not foolhardy. It is merely extraordinarily ambitious: there are twelve thousand recorded species of plants and animals that humans have eaten, and millions more varieties. Which ones do we preserve? Frankly it means discerning—sometimes just plain guessing at—the species and varieties that will be most important to human and ecosystem health.

Preserving seed diversity is a kind of pragmatism in an epoch of warming. It presumes a future where there is space to plant, laboratories in which to experiment, and greenhouses for growing, as well as people to harvest and eat the new varieties. To some, this emphasis on moving forward with practical, adaptive solutions seems to undermine carbon-mitigation efforts to reduce carbon. But it need not. Rather, the production of food (indeed, all manner of consumption) can be designed to work synergistically with carbon mitigation.

Generating food in a time of global warming occurs across vast spatial scales. The aim is simultaneously to produce good, nutritious food, decrease greenhouse gases, and adapt to a changing climate (in the face of declining oil reserves—what is known as "peak oil"—accelerating population, and declining availability of freshwater). These parameters describe a complex system with many interacting parts, ones that are usually invisible to the consumer. It is for this reason, among others, that people are choosing to reacquaint themselves with the food chain they are a part of and play a larger role in their own food production.

Pickup days at the Intervale Community Farm are Monday and Thursday. I chose Thursday so that the refrigerator would be full as the weekend came on, and shopping would be one less chore to fit in. With over two hundred members, the Intervale Community Farm is perhaps the most successful community-supported-agriculture enterprise in northern Vermont. Beginning in June and continuing through October, with a winter share available November through May, the farm provides all the vegetables a family of four needs, for $40 per week.

Before my own garden was established, my family and I purchased a share of the farm's warm-season produce. As in most northern gardens, the first foods to come off the land were greens—lettuce, spinach, and arugula. We ate these by the pound, refueling our bodies with fresh, green growth, vitamins, and chlorophyll after five months of winter. Radishes, asparagus, and spring turnips followed next, and in July the farm was full of carrots, raspberries, broccoli, and the first squash. By September we were waist-deep in produce, and the farm erected signs informing us of what we could pick on our own, no limits. We brought home cherry tomatoes by the bushel, which my children ate like Halloween candy, and veritable bushes of basil filled the car. Summer evenings were spent pickling beets, grinding pesto, and preparing the excess for the freezer. It was impossible to eat all the food the farm supplied for us.

The Community Supported Agriculture (CSA) model has grown since its origination in Europe and Japan in the 1960s. CSAs provide farmers with money up front, prior to the availability of produce, to assure farmer viability if weather catastrophes or other damaging events occur. Farmers who use the CSA model are able to focus on growing food during the season, as the market is already formed. Consumers share the risks and rewards with their farmer, and the community supports its working landscape to assure quality food, ecological stewardship, and the financial success of the grower. It is a statement of what the community values as it invests in "shares" of the farm's harvest.

The CSA model is expanding. Farmers include foods grown "off the farm," like bread, cheese, cider, grains, meats, and eggs. Some CSAs include fair trade goods that come considerably farther than the one-hundred-mile distance most locavores demand. At the Intervale Farm, coffee is sourced from a Nicaraguan whose farmers have formed a kind of sister-farm relationship with the Intervale, complete with reciprocal visits and conversations about building local markets amid the dominant exchange of globalization. Other farmers have begun to include a fuel share: wood pellets and cordwood are expanding food CSAs into local energy CSAs, adding to the consumer's motivation to preserve local farms and support farmers year round.

Look out onto the landscape where you live and imagine your watershed; high points of land collect water that flows downslope, feeding streams, brooks, creeks, and eventually rivers. A foodshed is similar: a geographic area from which a population's food flows. Foodsheds, like watersheds, are governed by geography. Their protection is elemental to the evolution of an alternative, local food system. When people rely on local food, they can connect what they eat with where they live, and in this way develop a sense of place and stewardship.

On a fall day, I prepare a meal that will serve as both lunch and dinner. As the seasonal weather comes on I cook in the late morning, and the house fills with savory smells. November has been warm, "warmer than average," the meteorologist reports, and surprisingly sunny. With clear skies and few clouds to hold in the day's heat, the nights are frosty but the days still feel like late summer. My garden remains productive.

On this particular day, I feel like making Portuguese soup. The screen door slams behind me as I stroll out to the garden and survey it. Most of my raised beds are growing a crop of winter rye. Others are strewn in straw, which makes a top dressing for the bulbs of garlic planted underneath. But some summer green leaves remain. One bed has a couple dozen leek bulbs standing solidly in the ground as their frond-like leaves brush the top of the soil.

Another bed is covered with a handmade cold frame. Two storm doors form the A of a triangle structure, and pieces of salvaged pine enclose the box at either end, keeping out the cold and the critters. I pry open the glass and put my hand in to test the temperature. The day's sun has brought it to a balmy 70 degrees Fahrenheit (21 degrees Celsius), fully twenty degrees warmer than outside. Our experiment of growing greens into winter is working. Thriving in this simple structure are baby arugula, winter lettuce (a variety known as Rouge d'Hiver, meaning "red of winter"), and spinach with quarter-sized leaves as thick as watercolor paper.

A final raised bed grows thick-stemmed, curly Russian kale. Planted in June, the kale is still vigorous and stands a foot tall. On frosty mornings it wilts and collapses, but as temperatures rise it regains its turgidity and is upright again. I stoop and gently break off ten stems.

I have a few small potatoes left over from my potato crop, which suffered an outbreak of late blight. I grab several onions and a bulb of garlic. All of it goes into a pot where I am sautéing chorizo from my friend Beth. She makes her spicy sausage from grass-fed beef and pastured pigs that my children and I know as calves and piglets in her barn.

Soon the smell of the soup circulates through the house, evoking a household from the past, when someone was always at home to cook. I return to my writing, infused with the serene feeling that comes with growing my own food and purchasing food from people I know personally.

In my mind I call certain actions "true to life" because they provide the opportunity to avoid faceless consumption and take in resources in a way that establishes relationship with others. Further, true-to-life actions encourage the living systems around us. In some small way, they are the antidote to modern depletion. To reshape our daily existence is to align with true-to-life endeavors and open the passageway into a new era of adaptation. The way we produce food, build buildings, and transport ourselves across terrain, these are actions we can control, ones that advance the concept of persistence.

All around me I see people practicing true-to-life actions, acknowledging that they themselves can be catalysts of change, even goodness. A true-to-life action carries the weight of truth. It is buying beets, turnips, and potatoes from the people who grew and stored them in a root cellar to be sold at a winter market. It is the sight of healthy children eating heirloom apples and zesty carrots in school lunchrooms. It is an industrial building with roofs covered in solar panels. It is a barn that disperses its wastewater across a constructed wetland where plants take up the valuable nutrients. True-to-life actions are designed to function like living systems: powered by the sun and cycling wastes. They are also deeply pragmatic.

There is a local-food revolution emerging in America. Farm stands and farmers' markets are multiplying, and community-supported agriculture is garnering a larger share of the food market every year. New patterns of purchasing and eating are emerging too. My own family is experimenting with eating only local foods: eggs and oatmeal for breakfast instead of packaged cereal; cider rather than frozen lemonade. By pledging to eat locally, first for a whole week, and then a whole month, we have joined a league of people who are ready to test old patterns of consumption. After a month's time, it is hard to buy bananas or reach for Chilean grapes, given the distance these foods have traveled. Like the transition to recycling, eating locally has established a moral component to my purchase of food that hasn't been there before. Food has become the medium through which I express my willingness to mitigate the carbon of my species. But beyond that, food is now an expression of my affinity to a landscape that demands participation, not everyone's, perhaps, but my own, mine personally, if it is to remain resilient, adaptive, and vibrant as the future unfolds.

Within a mere five years, the locavore movement has expanded beyond the imagination of even the most dedicated farmers and longtime local-food consumers. New magazines, eating clubs, restaurants, and websites have sprung up in support and celebration of the benefits of eating food grown within one hundred miles of home. As constraints are met, new ideas have materialized. For instance, Vermont has plenty of sugars (maple syrup, honey, even beets) and a dairy industry that struggles through, but the region has not traditionally been a big producer of grains, and without local wheat there could be no local pasta, bread, baked goods, or flour tortillas. It has abundant animal protein but little soy. Soon farmers began experimenting with wheat and barley, and hops for beer. Soy is now grown and fermented into tempeh and tofu, and waste from production is made into a non-toxic wood preservative. Chic restaurants now cater to discriminating diners who want a variety of in-season, locally procured foods.

What is so remarkable about the local-food sentiment is how adaptive it is to a time of warming. I think of it as a preadaptation, an evolutionary precursor to a future world where shipping food fifteen hundred miles (the current average) will be considered as outdated as a telegram.

~~~~~

In 2008 I attended a conference on agricultural adaptation to climate change at Teal Farm, a sustainability think-tank in northern Vermont. Dairy farmers, researchers, permaculturists, and climate activists got together to predict how the Northeast would change: temperatures will increase 4–7 degrees Fahrenheit (3–5 degrees Celsius) by 2100; the landscape will be characterized by more heat waves, heavier precipitation, and higher winds. Our region has already recorded a 10 percent increase in rain and snow since the 1970s, and it is anticipated to experience 30 percent more over the next century. Given these predictions, we gathered to discuss the practical question of growing food.

Agronomists at the Rodale Institute in Kutztown, Pennsylvania, have been experimenting with alternatives to conventional agriculture for thirty years. And they do it surrounded by industrial-sized row crops where corn is rotated with soybeans and wheat with little regard for the native ecology that once occupied the land. This is the kind of agriculture that dominates Middle America. Yet Rodale's approach has always been to work in the midst of the practices they were trying to change. And their emphasis on soil health, in particular optimizing the soil's water- and carbon-holding capacity, has won converts over the years.

Prairie soils once contained up to 20 percent carbon. They were black gold. Now the organic matter in conventional soils has declined to as little as 1 percent in some Midwest soils, equivalent to the poorest desert soils in the world. Therefore Rodale's mission to educate farmers on the merits of increasing soil carbon is particularly timely. Its researchers have found, for instance, that no-till farming, where farmers do not turn under the topsoil in preparation for sowing, significantly increases soil carbon and soil biodiversity. These factors buffer against climate extremes and expand the opportunity for cropland to sequester carbon.

Rodale's approach has been adopted by farmers across the country. But there are other models as well, ones that start from an ecological concept of what the landscape "wants" to become versus revising conventional practices to make them less bad.

My erstwhile student Ben Falk is a possibilist, and he evokes the same in his landscape clients. People who come to him realize the scope of the dilemma and embody an active hope. His clients are motivated to move beyond grief and judgment. They see the opportunity to respond positively given the land they own, and they want to launch themselves into the questions "What's possible here?" and "How do I adapt the landscape where I live to generate food?"

For gardens in the Northeast, Ben Falk begins with tree crops, nut and fruit trees in particular. Tree crops, like the forested ecosystems that have historically occupied New England's landscape, can be the backbone of an agricultural system, increasing tolerance to temperature and water fluctuations, and to disease. Tree crops also increase permanent soil cover; they stem nutrient leaching by holding on to minerals with their roots, and they improve the way nutrients cycle. Unlike annual agriculture, there is never bare ground with tree crops where nutrients can easily wash away. And they don't only provide food. Tree crops can generate energy through fuel wood, enhance microclimates and create windbreaks, and provide habitat for animals (wild and domestic). Like forests, they are multistoried elements: dense, often diverse, and self-replicating.

"Adaptation is, in fundamental ways, inherently local," writes Raymond Motha, an agronomist with the Department of Agriculture, "[and] our response should be tailored to local circumstances." Food production is an easy and appropriate entry into adapting to the Age of Warming and the uncertainty it brings. By purchasing food from people who grow it ecologically, we move toward a better fit with our environment, one that respects coexistence and promotes our own persistence at the same time.

CHAPTER 5

Our Oldest
and Newest Energy

To enter my basement you have to lift a heavy hatch, like entering the bulkhead of a ship. You put your finger and thumb through a thick metal ring and heave the door to a locked position. A short flight of stairs that resemble a tree house ladder leads down to the basement, where a single lightbulb hangs from the ceiling. Boxes of unused housewares stand in one corner: a beer-making kit, a rusty iron, a hand-crank ice-cream maker with a salt-encrusted barrel. There is a standard water heater too, customary in most basements. Five feet tall and cylindrical, it stands soldier-like in the corner, a testament to the modern conveniences that Americans have come to expect.

What is unusual about my basement is the bank of batteries to the left of the heater. There are twelve of them. At three feet tall and one hundred pounds each, they take up the majority of the room's small footprint. Above the battery bank and connected to it is a piece of electronics the size of a toaster oven. Red and green lights blink from tiny bulbs, and a digital display flashes a number. This is the inverter, a device that transforms the electrical current from the batteries into one that appliances can plug into. A charge controller with a gray ready-grip handle is within reach to shut everything off, just in case.

Homeowners don't expect their houses to come with a bank of batteries, an inverter, and charge controller. Neither did they expect their houses to be furnished with hot-water heaters in the beginning of the last century. It wasn't until the early 1900s that water heaters were introduced to the American household. Before then, hot water was generated in a kettle on the stove and carried between rooms: kitchen, bedroom, and bath. The notion of an appliance that produced continuous hot water would have seemed outlandish; it would even have run counter to the prevailing culture, as until the late 1800s bathing was considering effeminate and rather unhealthy.

The first water heaters were clumsy and imperfect. Some ran on coal but had no vents, so houses paid for their hot water with sooty air. One prototype, the Jewel, weighed forty pounds when empty. Made of cast iron, the portable device was the equivalent of carrying eight frying pans from kitchen to bath and back again. A mere fifty years later, hot-water heaters were de rigueur.

I began producing my own energy not because I was fascinated by engineering or because I had any mechanical expertise; rather, as a research ecologist stationed in far-flung outposts, I needed light, music, a computer, and some way to tell the world I was OK. In remote Alaska, we flew a generator into the bush along with tents, motors, and three months of supplies. The generators were table-sized machines that we trekked from airstrip to our island camp via an inflatable boat. On several round trips, we carried two fifty-five-gallon barrels of diesel, spilling a little here and there. The generator worked well enough. It took several tugs on the lawn-mower-type pull cord to get the carburetor started, but it produced enough energy to run our computers and ham radio and to light up our canvas tents so that they glowed a soft yellow among the sedge and moss muskeg. Once, ice from the Bering Sea drifted around us during a week-long storm and barred our escape by boat. We ran out of diesel and used the last charge of our radios to send an SOS. When the Coast Guard helicopter arrived, we were fine, just hungry and without power.

In the backwoods of Colorado, five miles from the nearest utility pole, I took advantage of three hundred days of annual sunny skies and used solar panels for the first time. Living in a cabin at high elevation in a county where 95 percent of the land is either publicly owned or in wilderness, I generated my own power. I had simple needs: one reading lamp, a laptop computer, and a boom box. They all ran on two 100-watt solar panels on the cabin's roof. Panel wires entered the log cabin through a hole drilled in the chinking and led down to two lead-acid car batteries with Sears labels still glued to the side. A small inverter transformed the current from batteries to an electric strip, and the entire setup for the summer was less than $1,000.

These were my introductions to self-generated power, and these systems are virtually unchanged after twenty years. My off-the-grid house on a wooded hillside in Vermont's Green Mountains gets its energy from the same set of parts. Making electricity from the sun is a technology discovered in 1959. Over the half century since, the technology has upped its efficiency from 10 percent to 28 percent. Photovoltaics rely on silicon wafers to store the copper wire that runs through them. When sunlight hits the copper wire, it excites electrons to move, which makes an electric current. The current is then stored in batteries. Like the Jewel water heater, photovoltaics have been updated. Our generator doesn't need a pull cord, and the batteries, while lead-acid, hold ten days of charge. But the principles are the same: generate electricity from the sun and store the excess in batteries for the times when the sun is not shining.

My house's roof and south-facing walls are completely covered in solar panels. They have been mounted to collect the rays of the sun once it rises above the maple, birch, and hemlock forest that bounds our property to the south. The arrays look like mirrored sails and cast a silvery iridescent color not unlike that given off by calm water on a summer's day. The aesthetic is beautiful and functional and the panels convey an unmistakable message: a modest American home can power itself with energy from its roof.

The sun is the primary source of energy for life on Earth. Almost every living thing is solar powered. Without the sun there would be practically no life of any kind. The leaves of plants are like solar panels, converting captured sunlight to power their photosynthesis and split water into hydrogen and oxygen. The free hydrogen then binds with other elements to grow leaves, stems, bark, and sugars and starches, the basis of nutrition for the primary consumers in the world's food chain. A plant's ability to transform photons of light energy into stored energy is wholly effective, a biological process that has evolved over three and a half billion years since the first photosynthetic organisms evolved from bacteria.

At five billion years, the sun is the oldest energy on earth. It is older than the oil in the Gulf of Mexico, older than the coal in West Virginia, older even than the wind that blows because of the sun. Not only is the sun older than other forms of energy, it is also more abundant. The amount of sunlight hitting the earth in a single hour is enough to power the entire world's economies for a year. This realization is startling, astounding, and breathtaking. And it forces us to ask: in the Age of Warming, how can we become a solar-powered civilization?

⁓

Vermont lies at latitude 43, just south of Germany, which, at 42 degrees latitude, is comparable in its annual solar gain. Like Burlington, Vermont, the city of Berlin is usually cloudy and receives on average only ninety days of sun each year. It is a rainy and snowy place with dark winters and, happily, sunnier summers. In contrast to southwestern states like New Mexico and Arizona, or the southern European countries of Portugal and Spain, Vermont and Germany have relatively meager solar resources. For every kilowatt of solar power installed in these sunnier places, Vermont and Germany need to install twice as many to get the same return. It is surprising, then, that Germany is a top manufacturer of solar panels in the world. Moreover, Germany has more solar energy installed per capita than anywhere else. This is something of a feat given that the nation of Germany had only a handful of installed solar panels before 1990. By 2006, half of the world's solar-energy technologies were produced there, and by 2009, one in every one hundred German homes had installed solar photovoltaics.

On the first clear day after a winter storm, I brush snow off my solar panels. For the lower set, a broom is long enough to sweep the snow away from the angled surface. If the snow is light, a simple firm tap at the base of a panel brings it off the entire array in one galumph. Reaching the roof panels is harder and demands a metal rake that I extend thirty feet into the air. I wrap a towel around the rake's rigid edge and hoist the clumsy weight to the upper part of the array. As the rake bends and tilts in all directions, I look as if I'm juggling a pile of spinning plates. At last the rake lands on the panels and with slight maneuvering from the ground, the snow is shed. Sunlight hits the radiating panels and my house immediately generates its own power again.

There are no two ways about it. The phenomenon of global warming demands that we replace fossil fuels with carbon-neutral energy. The world's electricity demand is set to double by 2050 and triple by 2100, so it is imperative that we make the transition soon. Not only will nations need to accommodate the increased electric use by those who only nominally have it now (the low-energy users with intermittent power), but they will need to provide it for the two billion who currently live without any electricity and the three billion who are yet to be born.

Replacing fossil fuel energy with alternatives is well under way. Municipalities, neighborhoods, and entire nations are looking to alternatives. Many, like Denmark, are investing in large-scale wind. Others are installing large-scale solar. In contrast, new nuclear plants are being proposed across the world even while moratoriums on nuclear have held for thirty years in the United States. (The last nuclear power plant built in the US was in the 1970s.) These large-scale projects are meant to duplicate the scope and centralized power generation that we now get from coal and natural gas plants. Huge central clean-energy plants could exchange a polluting technology like coal for a cleaner one like large-scale wind, sending the electricity thousands of miles along transmission lines that crisscross the country.

In 2008 the United States relied on coal for half of its electricity, most of which came from coal-fired facilities in Ohio, Pennsylvania, and Indiana. The coal itself comes from Appalachia—West Virginia and Kentucky, primarily. The pollution load from coal is so great that coal plants emit more than 80 percent of the carbon dioxide that comes from the entire electricity sector. In other words, when we look at per kilowatt hour of electricity produced from natural gas, wind, nuclear, solar, or coal, coal delivers well more pollution than all other sources combined.

People are beginning to see that organizing US electricity production around centralized coal plants is problematic, not only because of carbon emissions, but also because coal engenders a host of intractable environmental problems. Margaret Palmer, an aquatic ecologist, argued that mountaintop removal of coal should be halted altogether, at once. Palmer lists the litany of environmental and health impacts: toxic water, toxic soil, polluted air, species loss, mudslides, lung cancer, asthma, and chronic—often fatal—heart, lung, and kidney disease. "The science is so overwhelming that the only conclusion that one can reach is that mountaintop mining needs to be stopped," Palmer wrote.

So, in a time of warming, how do we wean ourselves from fossil fuels and transition to a low-carbon-energy world? Some promote natural gas as the intermediary between less efficient hydrocarbons like coal and petroleum and those that are zero emitting, like wind and solar. And it is true that there is no shortage of natural gas; according to the Department of Energy, the United States has approximately 200 trillion cubic feet of natural gas reserves, higher than elsewhere in the world except Russia and the Middle East. While natural gas can bridge a US transition from nonrenewable to renewable power, it certainly isn't without its difficulties. In the Rocky Mountain West, where large natural-gas deposits are extracted from coal-bed methane, the industry has contaminated freshwater aquifers. Similar issues are being raised in Pennsylvania and New York, where hydraulic fracturing of natural gas wells uses industrial chemicals and up to six million gallons of freshwater per well.

Solar thermal generation is currently the most economic when it comes to solar-energy technologies. Parabolic mirrors rapidly heat a circulating fluid to 250 degrees Fahrenheit (121 degrees Celsius). This creates steam, which can be used to generate electricity. Additionally, the hot liquid can be transferred to water for space heating. While concentrated solar arrays are normally built on an industrial scale, inventors are deriving ways to set up household versions, an idea as revolutionary as hot water flowing from a kitchen tap in 1910. Photovoltaic technology is also expected to make great economic breakthroughs in the coming decade. Spectrolab, a subsidiary of Boeing, projects that photovoltaic cells will nearly double their efficiency by 2013, up from the average 20 percent efficiency that commercially available panels are now limited to.

Solar thermal, however, takes up a huge amount of land. (A 350-megawatt solar thermal plant in California's Mojave Desert occupies one thousand acres.) That said, there are many companies and university research programs pursuing lower-cost approaches to photovoltaic cells to seamlessly integrate them into building materials, using a product known as thin-film cells. This would replace conventional and rather bulky panels that are installed postconstruction. Analysts envision that integrating photovoltaics into the siding, roofs, and perhaps windows of buildings will drive down the price of solar power. Additionally, the installation of large concentrated solar photovoltaics by utility companies will lower the technology's cost, helping to bring down the price of solar to $1 per watt, in line with wind and coal.

Whenever I consider that Earth receives 100,000 terawatts of solar power at its surface each hour, I am astounded. To put this in perspective, 1 terawatt is equal to 1 trillion watts. The human population uses approximately 16 terawatts of energy each year. Therefore, what arrives hourly from the sun is nearly six thousand times more energy than the planet's population consumes yearly.

We have an unlimited supply of energy, and it is our challenge to capture it the way an unfathomable number of plant cells currently do. As investment in solar technologies increases, coupled with economic incentives and decreasing costs, residential and commercial space is being "mined" for its solar capacity. Large flat-roofed megastores, from Wal-Mart to Whole Foods groceries, are installing panels to offset electricity costs and deflect heat away from buildings. Military bases are putting up solar on airplane hangars, and FedEx has covered its New Jersey hub with panels.

A technological breakthrough in solar is more imminent than one in coal. This is true not only because carbon capture from the 614 coal plants in America would require the transport of twenty million barrels of liquefied carbon dioxide to underground storage each day, but because solar is a nimble technology. Similar to the way a mite hitchhikes on the beak of a hummingbird, solar can use the built environment around us.

The sun not only supplies Earth with solar power, it also generates wind. Differences in air pressure are due to the response of Earth's surface to heat from the sun; high pressure results from cooler, dry air, and low pressure from warm, wet air. Molecules of air therefore are moving in response to the sun. When we generate wind power, we are in essence collecting another form of solar energy.

Wind power is advancing faster than all other renewable and nonrenewable energy sectors, with the exception of natural gas. After decades of research on turbine design, wind power is now economically competitive with coal and costs between five and seven cents per kilowatt to make. Although wind is intermittent, when it is coupled with solar this intermittency declines drastically. The reason for this lies in the fact that the wind blows more often at night, when photovoltaics are naturally suspended, and during the winter and spring months, when the skies are cloudier.

In 2008 the US Department of Energy released a report claiming that wind generation in the United States could satisfy 20 percent of our electric needs by 2030 if installation increased threefold by 2017, and a direct-current transmission infrastructure was built to connect the wind in the Midwest, namely the Dakotas, Nebraska, and Wyoming, to high-use populations. A regional grid such as the one proposed would allow slack in the system; the wind does not need to blow everywhere at once, only in different places at different times.

Wind is an alternative to fossil-fuel-based power, one that has proved itself on the margins, is maturing swiftly, and is moving to the mainstream. But unlike solar it is rarely suited to a person's home. This is where community energy comes in.

In 2009 the Institute for Local Self-Reliance, a policy think-tank in Minnesota, published a report titled *Energy Self-Reliant States.* In it, John Farrell and David Morris advocate for state-based production of renewable energy. They limited their analysis of power sources to small onshore and offshore wind farms, rooftop solar, small hydroelectric, combined heat and power (where biomass is used to produce heat first and electricity second), and conventional geothermal, which uses steam heat close to Earth's surface. These limitations allowed them to test whether dispersed generation is as viable a substitute to carbon utilities like coal as vast wind projects seen in Europe. Farrell and Morris reasoned that smaller projects would mean fewer transmission lines, localized power generation, and greater economic benefit for the states.

Farrell and Morris report that 64 percent of US states could generate *all of their own power* with renewables. Another 14 percent could generate 75 percent of their electricity with renewable biofuels from grass, algae, and waste methane. Farrell and Morris maintain that decentralized energy production, where electricity comes from communities, neighborhoods, and municipalities rather than centralized plants several hundred or thousands of miles away, is the most economical and sustainable vision of a fossil-free future. Why ship electrons from Ohio to Massachusetts and between Missouri and New Jersey when Massachusetts and New Jersey can produce their own?

"Small is beautiful," wrote the economist E. F. Schumacher. Like local food, local energy generation is a way to build resilience into social and economic infrastructures. Energy, after all, is the currency of our economy and without it we will have no commerce. Yet, as demonstrated by my empty barrels of diesel in the Alaskan outback, we thwart all resilience if we rely on fuels that are igniting crises.

Montana is a windy state, as are the Dakotas. The wind that blows there could produce far more electricity than each state currently needs, one hundred times more. But they are not alone in this distinction. In reality, most western and midwestern states could generate all of their electricity with onshore wind alone, a fact that is not true of eastern and southeastern states, with Vermont and Maine being the exceptions.

Half of California and Nevada's energy needs could be supplied through rooftop solar. And in Alaska, the Land of the Midnight Sun, 20 percent of its electricity could be generated from rooftops. But these are very conservative estimates, because Farrell and Morris base them on rooftops alone. They exclude parking lots (950 square miles exist in the United States) and ground-mounted systems where solar panels are installed on poles adjacent to buildings. They also exclude marginal lands, highway right-of-ways, and solar electric technologies other than photovoltaic. If these approaches are included, solar could contribute significantly more to the energy generation in each state. For instance, if 10 percent of the nation's highway right-of-ways were developed for their solar potential, they would generate 100 percent of the country's electricity needs: 100 percent from roadside solar alone.

Farrell and Morris's point is this: we do not have to rely on single, large-scale sources of energy, even if they are renewable. Instead, integrating different types of nuclear-free and carbon-free energies mitigates carbon, optimizes efficiency, diminishes transmission distances, builds economies, drives the transition from fossil-fuel electricity to renewables, and increases our resilience to the surprise events that will characterize the Age of Warming. Presently, a shock to our electric infrastructure can plunge entire regions into the dark. Under a decentralized and renewable-energy infrastructure, a shock would take out relatively few lights.

On a late January evening, I hear what sounds like a snowplow coming into the driveway. It is eight o'clock, and we have been home from school and work for three hours. The computer has been on, and so has the stereo. I have baked a casserole in the gas oven, an appliance that uses an electric glow plug to stay lit. My daughter's request for a clean pair of jeans prompted me to put in a load of laundry. And all along I have been running water in the sink, using the cordless phone, and turning on lights.

I dread the sound that is coming from the yard, from a shed, actually, that sits to one side of the house. It encloses an 8,000-watt propane generator that automatically detects when the batteries in the basement need charging. The winter sun is weak and hangs low in the sky. Truth is I haven't seen it for several days. This time of year, it, combined with the limited number of panels on the roof, is not enough to electrify my twelve-hundred-square-foot household for my family of four.

Living in a renewable-energy home with power that fluctuates over the year, we are more in tune with our usage than most families. Our place carries all the modern conveniences of an American house. We vacuum our rugs, have high-speed Internet, and a stereo plays virtually nonstop. Like other homes, we have a power-charging station on a household desk that catches mail, flyers, and the latest newsletters from school. Here iPods, iPhones, pagers, cameras, and rechargeable batteries are lined up and plugged in to a wall socket nearby. A handheld mobile phone, charging in its cradle twenty-three hours of the day, allows us to talk while we walk. And Christmas lights, strung across the porch railings and entrance eave, brighten our home mid-December through February. Our house hosts three computers, a printer, and a dehumidifier. The kitchen is fully equipped with a state-of-the-art refrigerator, toaster, and all manner of blenders, choppers, and other culinary devices that require plugging in. We have a chest freezer that we fill with produce gleaned from the growing season, and my husband is partial to freshly ironed shirts, an energy-intensive whim but one we accommodate. In the garage we have the requisite power tools, including circular saw, drills, and wet-dry vacuum. Our weed whacker and lawnmower are both electric appliances with long cords. Mowing on a summer day, we get the unusual pleasure of using the sun's energy to cut grass.

While there are few devices we do without, living off the grid, we are judicious in our use of power. Few of our appliances remain in "ready" mode, their digital displays blinking, drawing power when turned off. And while we have a clothes dryer, we rarely use it. When the school held a fund-raiser selling compact fluorescents by the four-pack, we bought several and replaced all our traditional lightbulbs. We admonish our children to shut lights off when they venture between rooms, and it is seldom that the outdoor Christmas lights remain on through the night.

Because we do not have a utility line or meter running to our house, we keep track of the amount of power in the system ourselves. A digital display, tucked into the laundry closet, tells us how much the solar panels are generating, how much the house is using, and how well charged the batteries are. There's a diode that turns green when more power is coming into the house than we are using. Even better, when the system can no longer accept energy because the batteries are full and all power loads are met, the display reads FULL. Those are fortunate days, ones that begin in April and extend through October. But they rarely describe the month of January.

It was a hasty purchase when my husband and I bought our place in 1997. In the late 1990s in Vermont it was a seller's market, and the day we drove the Hollow

Road to look at what had been advertised as a "Starter Home in the Woods," there was a stream of cars lined up before us. Couples peered in the windows of the six-hundred-square-foot camp, assessing the quality of light, the pine floors, the ship's ladder to the sleeping loft. It was a modest place on ten acres, snug, heated with a woodstove, and quiet. A winter view of the Adirondacks captivated us. The cabin was off the grid, but there were no solar panels. Instead, a wind turbine topped a one-hundred-foot tower out back.

At the time, we had no experience with wind generation, though we'd seen wind farms as we drove through Wyoming's Red Desert. Against the expanse of the High Plains, the turbines appeared relatively small and sculptural from a distance. As travelers on the interstate, we saw them a mile away but passed them in less than a minute's time.

In Vermont our wind tower and turbine were legions smaller, but the principles were the same; above the trees, the turbine caught the laminar flow of wind, and a vane oriented the turbine to face into it. Electricity ran through the center of the lattice tower on heavy wire, then went underground and into the house, where it charged batteries. The machine was quiet, less than 1 kilowatt in size, and its light pine blades were hardly three feet across. After we bought the place and settled in to raise two daughters, the girls would find the shadow the turbine cast in the yard and stand at its center, blades and sun spinning around them.

Winds are strongest in the winter and early spring, the same seasons that solar gain is at its minimum. One blustery evening, I check the meter. The diode is green and the system is charging. We are converting something our neighbors complain about—high winds on a winter night—to a tangible good, a valuable commodity. Like picking blueberries hand over hand from a plentiful patch and then freezing the excess, storing excess wind power makes me feel secure, and smart. I am benefiting from nature's bounty without any ill effects.

Living in a solar- and wind-powered house has its advantages. One, the power never goes out. In August 2003, when overgrown trees struck a high-voltage line in Ohio and the eastern seaboard and parts of the Midwest experienced a blackout for several days, we had power. Or during the wettest October on record, when hurricanes were racing northward from the Gulf of Mexico, and trees, their roots waterlogged with rain, came down all around us, we had power. Yet having power when others do not is only one of our advantages. On a bright day, when sun fills the house and the sky is as blue as delphinium, I feel an inner contentment not unlike when friends gather around a table laden with food. The feeling is akin to the hunter who, having taken a deer, fills her freezer with venison and gives the rest away, or the builder who discovers a spring above his home and pipes the water downhill using only gravity. Generating electricity from photovoltaic panels, immobile objects that magically take light and turn it into electricity, exposes the wonder of physics. It feels like something miraculous is being unveiled.

Daniel Nocera is a wizened character. His inquisitive demeanor personifies the years spent in chemistry and physics labs, studying catalysts. When he talks, it is with a kind of humorous gravity. He makes grand claims and easily shows his displeasure when on a panel with people who have different opinions. I met Nocera at a conference on energy and the environment, one that asked out-of-the-box thinkers to speak but was sponsored by the largest energy and car industries in the country. At the start of the conference, Nocera was asked to present his "big idea." He had five minutes to turn the conference goers on to a new way of thinking, a glimpse into a paradigm shift that he was helping to orchestrate.

Nocera's big idea, the one he's been working on for more than a decade in his lab at the Massachusetts Institute of Technology, is this: electricity equals plants. Photosynthesis, a technology several billion years old, replicated in the simplest of algae and the most complicated of angiosperms, makes electricity by capturing photons from the sun. Leaves, bark, algae, the moss growing on sloth fur, even diatoms living five miles below the ocean's surface, in the darkest, deepest space, are making electricity. Moreover, they are using it to split hydrogen from water in order to convert the hydrogen into glucose, starch, and sugar molecules that fuel their many life-history tasks: grow, flower, set seed, and, if perennial, survive dormancy.

In Nocera's vision of the future, we mimic a plant's ability to make energy from the sun, copy its properties, and replicate them household by household. If an average American home uses 20 kilowatt hours a day, then, like a photosynthesizing plant, each household would need to split five and a half liters of water a day. Six plastic liter bottles of water, Nocera says, contain the energy for a home.

Splitting water is an energy-intensive activity because hydrogen binds strongly with oxygen. While scientists have separated hydrogen from oxygen in water for decades, the process uses a lot of power, harsh solutions, and rare and expensive catalysts, like platinum, to do it. Nocera wanted to find alternatives. His goal was to use power from the sun and break down water under benign conditions, that is, room temperature and neutral pH. Further, he wanted to use a catalyst that was abundant, cheap, and easy to manufacture.

Plant catalysts are unstable, and therefore difficult to make. Plants still have the patent on that! Light is collected and converted into a current that is fed to oxygen-evolving-complex (OEC) sites in the plant, where oxygen splitting occurs. But in 2009, Nocera reached a breakthrough. He had focused on the platinum group of metals: rhodium, ruthenium, palladium, osmium, iridium, and platinum, a group clustered in the center of the periodic table. All these metals have similar chemical properties, including the ability to catalyze reactions and resist corrosion. Nocera had experimented with each, but nothing worked. Then, he says, he looked up. There, above rhodium in the table, was inexpensive cobalt. Working with graduate students, Nocera synthesized a cobalt-based catalyst that, in the presence of electricity, split water into its elements of hydrogen and oxygen. At first, the implications of Nocera's discovery may be hard to see. But what Nocera has done is solve a crucial problem in the renewable-energy transition: the ability to store the sun's energy in another equally renewable resource—water.

July is Vermont's sunniest month. The garden swells with growth. Squash blossoms flower far beyond what the plant could ever make in fruit. Giant sunflowers grow from a single dime-sized seed to a five-foot-tall leafy plant; their stalks are fibrous and lignin-rich. Apple branches give way to young apples that hang in clusters of three and five, each a plump encasement for the dispersal of the caramel-colored seeds growing within. Overnight, a morning glory, barely past shoot stage, wraps its way up the stick fence, and slender grape vines spiral off rather than remain tethered to a wire.

The sun is high in the sky, lighting the garden by eight in the morning. It remains well above the yard's forest boundary much of the day. There is practically no shade. By noon the solar panels have been making electricity for four hours, filling the batteries. Inside the house, we've used relatively little power, having no need for lights and little interest in baking. The water pump has come on and the computer, radio, and Internet, of course. The green diode on the energy meter is blinking, and the meter reads FULL. As much as I dread the sound of the propane generator in winter, I am equally frustrated by our inability to store July's abundant sun for later use.

The size of our renewable-energy system is based on a series of compromises: cost, amount of house and roof area that is south-facing, efficiency of available technologies, and fluctuating radiance of sun. These all affect how large our system is. If cost were not limiting, we'd buy twice the number of batteries; we'd double our solar array and put a 3-kilowatt turbine on our wind tower. This upgrade would ensure a final silencing of the generator, but it would cost tens of thousands of dollars beyond what we have already spent. Plus, in July, we'd be so flush with power that by 10:00 a.m. the system would shut down.

That is why Nocera's research is so captivating. In effect, our household is an illustration of the energy transition yet to occur in the United States. We have abundant renewables much of the year, but an insufficient way to store the excess that could then be used in lean months. Further, we are reluctantly accepting technology (residential solar panels have only gained 10 percent efficiency since 1980, and lead-acid batteries, invented in 1859, have changed little in principle since then) in the context of a twenty-first-century phenomenon that demands technologies suited to its scale and urgency. We are waiting for the mechanical arts to catch up. Installing Nocera's water electrolysis in our basement would allow us to generate hydrogen in July to feed a hydrogen fuel cell—an advanced technology whose only emission is water—in January.

In this transformative reality, a household's energy would be generated and managed by the individuals living there. Larger consumers would need larger systems. Individuals who were conservative about power or who had invested in energy efficiency would use less. Home energy systems could be like hot-water heaters now; they come with the house.

In my ideal vision of a renewable-energy future, I manage not only my household electricity use but also budget electric power for a plug-in electric car. At some point in the near future I hope to absorb some of July's abundant sun, and travel forty, fifty, even one hundred miles on electrons that are now prohibited from entering my energy system because it is FULL. This ideal may not be far off. In a design coined "Solar Trees," the software company Google has covered parking lots with solar-panel roofs and attached outlets. Moreover, it has experimented with plug-in hybrid electric vehicles (PHEV) in an attempt to integrate them with solar generation. Nine PHEVs owned by the company have been tested on California roads and under California sunshine to determine how successful the venture could be.

PHEVs are different from regular hybrid electric vehicles. They have higher battery capacity, hold more charge, and use the electric motor for the first twenty to forty miles before transitioning to the combustion engine. Because 70 percent of Americans travel less than thirty miles each day, driving a plug-in car negates the use of a combustion engine, and its carbon emissions, most of the time. For me that includes two trips to the bus stop, a swing by the post office and general store, and occasional trips to a larger, nearby town for groceries and hardware, and to the library and music lessons: the usual kind of trips for an American parent who works from home.

In 2010 Google released the data on how well its nine PHEVs had performed. The Ford Escape Plug-In Hybrid, a prototype set to be released in 2012, could travel an average of forty-nine miles per gallon, ten or more miles per gallon higher than the Ford Escape Hybrid currently gets. But the Toyota Prius Plug-in Hybrid got ninety-three miles per gallon, almost twice its standard Prius (which Google also drove and recorded a forty-eight-miles-per-gallon fuel efficiency).

Researchers at the University of Vermont's Transportation Research Center, one of six federally funded research centers that focus on mobility, report that under the state's current electricity generation, Vermont could handle 200,000 electric vehicles, or a third of all registered cars. By plugging in at night, car owners take advantage of the abundant power available between 12:00 a.m. and 4:00 a.m., and utilities sell off-peak electricity they otherwise report as a loss.

This line of thinking has innovators realizing how electric cars and batteries can be storage units in the grid in what is referred to as "vehicle to grid" technology. Most cars sit parked twenty-three of twenty-four hours a day—often in the sun. Software engineers are incorporating battery storage in the design of optimal grids, especially as the grid encounters more-distributed generation from renewables. Soon, car batteries will be viewed as a way to power one's car and store household electricity, in the same way a tree's roots supply energy for spring's leaves or the breast cavity in a songbird bulges with fat, powering its flight southward.

Adapting to climate change is as much about discovering green technologies that use relatively little carbon as it is about changing behavior. Americans are notorious for using more energy per capita than most of the world. Unfortunately this isn't a recent statistic. Even in the nineteenth century we used nearly twice the energy per capita than elsewhere in the world. Now we use ten times. Is this the price of being American? It does not have to be. People are waking up to the idea that they can invent a lifestyle that is contemporary and rich and powered by sunlight. We will begin by conserving the energy we use from carbon-polluting technologies, drawing on nineteenth-century approaches to help us mitigate carbon rapidly. But all the while we will be innovating twenty-first-century technologies that are in keeping with the principle of powering ourselves with the sun, an adaptive strategy and a way forward.

CHAPTER 6

Localizing Home

During my childhood, our family-sized Coleman tent—sun-washed blue with net pockets for a compass or canteen—was the substitute for our suburban home on a quiet street. Tenting was the only time the seven of us slept together; my parents took the space by the door while we children were rolled out hotdog-style deep in the tent. A gas lantern hung from the center, and the older kids held bent paperbacks above their heads, reading in the last bit of light. Compared to our house, the tent felt safer, more intimate, like true shelter.

In my early twenties, I bought my first tent. At less than five pounds in weight, the Sierra Designs Flashlight comfortably accommodated one person and was suitably snug for two. It had exterior poles that clipped to the tent body rather than sleeves that you fed the poles through, which gave it a sleek, rocket-like look.

The Flashlight was my castle. Well before I owned a car or rented an apartment, I spent several summers living in the Flashlight with my backpack hung from a branch nearby. I cooked on a single burner with a collapsible cook set, each pot stored neatly inside the other. A Swiss Army knife and "spork" (a fork/spoon utensil combination) satisfied my cutting and eating needs. Richness was two pairs of clean socks.

The beauty of a tent is its confined space. There's only so much room, a feature that forces the camper to get quickly to basics: a long book, tinned fish, a wool hat, and camp mug. Versatility is the filter against which everything carried is tested. Pants that become shorts, socks that could be mittens, and a tarp that served as tent fly, ground cloth, or impromptu sled over a snow field. All were added as multipurpose goods. Small-boat owners will concur, and though I have met sailors who built a boat around a standup piano (so as to play original scores accompanied by whale via hydrophone), most sailors are parsimonious to a fault, taking aboard only the strict necessities.

In a tent, thin ripstop nylon is all that separates me from the elements. This is the other beauty of tent living: I may as well be outside. I smell the wintergreen crushed beneath me. I wake up to the sound of the wren. It is perched on the tent pole, mistaking a grommet for a hole to investigate. I hear the rain as it crosses the valley. The smell is musty and thick like basement air, and then *pat, patta, pat,* the raindrops hit the fly and I dig deeper into my tent-nest.

Tenting breaks down the walls that usually separate me from the story of life that, while continual in the animate world, is veiled as I move from building to building. When I step out of my tent in the cool morning or after a heavy midday nap, I immediately see the signs of visitors. A raccoon has patrolled the campsite, attracted by the fishy aroma from last night's dinner. It has pawed the ground for grains of rice and a fleck of tuna. Gray jays, also known as camp jays, swoop to the closest branches. I heat water for tea and drink it on a smooth rock that doubles as writing desk, dining room table, and stool. An ant scurries with a crumb in its mandibles. My makeshift, transient home forces me to shed my insulation and see that I am very much a part of the natural world.

My rural Vermont home began as a camp, a place for people living in the city to escape to. At six hundred square feet, it was well beyond a tent but too small for year-round living, or so the previous owners thought. The couple had built it themselves and, not being carpenters, had failed to make much of it square: the porch tilted to one side, and the seams in the drywall had split. A ship's ladder led to a loft where our double bed and narrow desk took up all the space. Downstairs there was an alcove for eating, a tight alley for cooking, and a ten-by-twelve-foot space for living. It was simple and comfortable. Friends used the term "cozy."

With my first pregnancy, we were compelled to build an addition. My growing belly bumped against the steep ladder, and I couldn't imagine taking an infant up and down the nearly vertical stairs. (Although we did consider a bassinet on pulleys.) Rough sketches and meetings with the bank and builder resulted in a twelve-hundred-square-foot house one year later. Suddenly the walls were thicker, the place grander, and there was no ladder. We had a second bathroom, a second bedroom, and what seemed like acres to live in. This change was perhaps inevitable. Like other young couples, we traded the space for the financial risk, signed the papers, and committed ourselves to thirty years to pay it off.

My house is not "green": it was not designed by an architect utilizing the principles of resource efficiency; it has no LEED (Leadership in Energy and Environmental Design) certification. Yet it is modestly well insulated, and we have replaced the single-paned windows with more efficient double-paned ones. We heat with wood gathered from the ten acres of forest that surround us or from a farmer who sells cordwood in our small town, and some of the lumber was cut from our land. But the most prominent environmentally responsible feature about my house is its size.

Like other middle-class homeowners who are aware of climate change, we have a list of green renovations we would like to make. One is to install a standing-seam metal roof in exchange for the asphalt, petroleum-based shingles. Second is to install solar hot water for our domestic needs, a renewable energy with a short payback period. Third is to blow in new insulation where the pink fiberglass has become compressed by resident mice. While we strive to green our stick-frame smallish home in the woods, others are striving to revolutionize green building from the start.

In 2006 the Cascadia Region Green Building Council (a group that advises builders in Oregon, Washington, Alaska, and British Columbia) began the Living Building Challenge. The challenge rates buildings on six performance areas, what they call "petals," with a goal of net zero impact on the energy grid, water systems, and natural environment that surrounds the building. For instance, a living building must produce more energy in a given year than it consumes. The water leaving the building must be as clean as the water entering. These kinds of design constraints have led to architectural innovation that goes well beyond making buildings green or sustainable, descriptors that define themselves in relation to conventional practices.

The design constraints in the Living Building Challenge force architects and builders to think of buildings as organisms, as components of the ecological world where they are built. How do they contribute to the health of ecosystems rather than being benign or, worse, degrading? Living buildings do not allow for the assumption of negative effect. Rather they strive to make the structures conducive to life: roofs that are habitat for winged creatures, constructed wetlands that receive wastewater and are planted with hyacinth, cattail, and blue flag iris, species that take up stream pollutants such as excess phosphorous and nitrogen and use them as nutrients to grow taller stems, deeper roots, and more flowers. In effect, the Living Building Challenge moves the conversation of human habitation into the arena of environmental ethics. How can human action be bent toward supporting life?

⟞⟝

In ecology, the term *local adaptation* refers to organisms that adapt to their geographic location, often at the micro scale. This idea contrasts with the notion that there are "type" specimens in "type" habitats. That is, each species can be described by a kind of average phenotype that exists across its range. Local adaptation, on the other hand, describes distinct microevolutionary adjustments that organisms have made in order to be extremely well suited to their habitat.

As a case in point, the pine leaf scale insect lives on conifers, their needles in particular. In the insect order Homoptera—meaning "uniform wings" for the two sets of wings that are similar in texture—they suck the juice from pine needles with pin-like appendages that extend from their mouths, a kind of insect fang. According to Don Alstad, a researcher at the University of Minnesota, pine scale has locally adapted to *individual* yellow pine trees in the Southwest, sometimes to *individual* branches.

Pine leaf scale insects are stationary most of their lives. The pine tree where an individual's mother lays her eggs is de facto her offsprings' home. To remain attached, pine scale secrete a protective layer of resinous glue to cover their bodies while they feed. Their conifer branch is the entire habitat they need: it provides shelter and food. And the following spring, when males sprout legs and wings and crawl out from under their cover to find a female, it becomes the place to reproduce as well.

What ecologists are able to show with case studies of local adaptation is how tight the relationship between organisms and their environment can be. Butterflies, for instance, are locally adapted to different plants to better couple their life cycles with the timing of flowers or the appearance of leaves. Similarly, fly midges will specialize on individuals of the seaside shrub called oxeye and become locally adapted to the range of mild but toxic compounds they produce. But neither butterfly nor midge is precluded from mating with individuals distant from the locales where they've adapted. In fact, the evolution of species to particular habitats maximizes the use of the landscape, dividing and redividing niches. In time, a geographic mosaic arises of populations intimately suited to their surroundings.

This concept of local adaptation as taken from ecology is being interpreted through the design of buildings, communities, even entire cities. As architects and civil planners grasp the climate forecast, they are incorporating local adaptation into their building designs. The challenge is what the architect Bill Reed refers to as "participating in nature," buildings that express a dynamic response to changing conditions the way living systems do. "Slowing down [environmental] degradation, while essential, is insufficient. Regenerating the evolving resiliency and matrix of life in each place is the other half of achieving a sustainable condition. We need to shift the purpose of design and process this way . . . embracing a whole-systems mind and design process to wholly participate in the system of life," Reed attests. In effect, Reed believes we can not only restore ecological systems but have our buildings—extensions of ourselves—play ecological roles in the natural communities they are a part of. We become agents that support living systems when we green the roofs of factories and provide nesting sites for meadowlarks or when we design constructed wetlands that not only break down our wastewater but serve as habitat for marsh-loving species.

The Willow School in Gladstone, New Jersey, is one such place. Designed in part by Bill Reed's architectural firm Regenesis, the school's trustees and faculty sought to make a locally adapted building and promote one of its three academic objectives—ecological literacy for students aged six through thirteen.

The collaborative that built the Willow School included landscape and building architects, environmental planners and site designers, as well as the school's teachers, who developed the ecological curriculum. By designing the thirty-five-acre site in unison, and in keeping with Reed's approach, the school and grounds became a living classroom and a model of regenerative design.

Constructed wetlands and outdoor gardens receive wastewater and runoff, ecologically treating polluted water. Permeable paving stones and parking lots, living roofs and bioswales (meandering ditches that retain storm water long enough to trap pollutants and silt in their vegetated slopes) all contribute to "fitting" the Willow School into its environment. Aware that the landscape used to be a forest before the school purchased it, the design actively encourages the land to become a forest again. In this way it will achieve its highest ecological potential: a living watershed able to store carbon, filter water, release oxygen, and house biodiversity.

Projects like these are not limited to the scale of a household or a community school. HOK, one of the world's largest architectural firms, is designing an eight-thousand-acre city in Lavasa, India, using similar principles. Known for its biomimetic designs—ones that mimic natural materials and processes—including an Asian skyscraper whose template is the structurally sound honeycomb, HOK is building a city that simultaneously restores the native ecosystem—returning an arid landscape abused by agriculture to a moist deciduous forest—while designing rooftops to trap, clean, and release water the way native banyan tree leaves always have. Even the insects of Lavasa have contributed to HOK's design. Similar to the way local harvester ants dig shallow channels around their nests to divert water, HOK has included a series of waterways in its master plan which will divert the monsoon rains from buildings. Biomimetic designs like these are not only adaptive, they collaborate with nature.

⁓⁓⁓

Phoebes nest in the corner eave above my office. In spring, the pair perches on the clothesline, bobbing on it as if it was a tightrope and they were its acrobats. I see them carry moss to their cubby nest and then return with heavy threads of burlap and strands of hair that fell to the ground after I gave my nephew a haircut outside.

I watch. The phoebes' bodies are in near-constant movement, tails flicking, heads cocking, eyes twitching, bodies searching. The black flies emerge and they catch them in midair, becoming my accomplices in keeping the stinging insects at bay.

A tiger swallowtail flies frequently past the window, beelining for the chive plants and nectaring deeply on the abundant purple flowers. She is also laying eggs, on fennel and wild parsnip at the perimeter of the yard. The young lilacs, in their second year of bloom, attract dozens of red admiral butterflies. My five-year-old daughter, Helen, makes a bluebird box in kindergarten and brings it home with instructions. "The bluebirds are in danger," she reports. We nail the box to a cedar post in the yard and she frequently checks the inside to see if someone is home.

But we do not always benefit from the life surrounding our house. There are carpenter ants swarming, and we see enough winged females to wonder if they are in our walls and tunneling in our wooden beams. Chip-

munks get into the stovepipe and land, sooty and pan-
icked, in the ash box. Raccoons topple the compost, spilling fruit rinds and coffee grounds. Japanese beetles eat
grape leaves, while curculio beetles destroy apples, bury-
ing into the flesh and spoiling them. And mice are ever
present. Sometimes I see their black peppercorn drop-
pings on the children's beds and wonder if they scurry
in at night, looking for crumbs or the sticky bits left on
candy wrappers.

Still, I strive for the ideal of making my home, one
already built and embodying the energy of its construction (spruce beams, gypsum, metal wiring, copper pipes,
fixtures, sockets, and wooden trim), into one that blends
with the natural world and restores in some way the natural area that existed before the house was built. Consequently, I have two bright blue barrels in the driveway.

Because the Northeast will experience greater deluges and more extreme droughts in this warming climate, storing rain in rain barrels for gardens and other
nonpotable needs is logical enough. My research began
online. With a simple search, numerous varieties popped
up. Simple blue plastic rain barrels with screens to keep
debris out were on one side of the spectrum. Elaborate
painted wooden barrels with copper faucets and a place
to attach a hose were on the other. The cost spanned
from $80 to $200. Most were composed of cheap plastic
with flimsy attachments, and few considered the details
of mosquito prevention and water overflow, given that
a quarter of an inch of rainfall on a typical household
roof can amount to two hundred gallons of water. Put off
by the cost, and because it seemed like a simple enough
project, my husband and I decided to build our own.

YouTube is a boon to the do-it-yourself (DIY) culture that is emerging in America. Anyone can post on any project, and building a rainwater-collection system is no exception. Lowe's, the chain hardware store, offered a five-minute how-to video complete with supply list. (Then a viewer did the math and found the supplies cost $283, far more than a basic rain barrel would cost. Would-be buyers beware.)

But there were other, noncorporate DIY-ers eager to share their experiments. "Mr. Native Texan" shared a seven-minute video of his "manifold catchment system." Here, four fifty-five-gallon barrels from a Coca-Cola bottle distributer were lined up beneath a rain gutter. Simple fittings connected the barrels to one another so that they filled simultaneously, effectively dealing with the water-overflow problem. A spout from one of the barrels led to a hose that ran downhill, and gravity delivered the water to the garden.

Watching the videos, I saw DIY culture at its best: innovation using materials at hand. There was Bob, a man of about thirty, collecting rainwater from a greenhouse he built. On camera he quickly sketches the design for his system, then fashions things he already owns to suit his needs: a piece of vacuum hosing drains the water from the gutter to the barrel, an old garden hose connects the barrel to his garden, a plastic flower pot peppered with pencil-width holes strains the water of sticks and leaf debris. Total cost: $40.

DIY culture in this mode is helping people adapt in place. Why purchase new fossil-fuel-derived plastics to build a rain-collection system when the reason for needing one is our dependence on fossil fuel? People are finding ways to break their dependence. Starting with their homes, they are realizing that carbon pollution is unacceptable. Like Londoners who banned coal fires after the Great Smog killed twelve thousand people in 1952, we are developing a new standard of what is appropriate behavior in the Age of Warming. In time we will reflect on the filth implicit in running a society on fossil fuels. But for now, we are inventing new ways to run our households.

One afternoon I head to a local brewer that I'd heard leaves its cast-off barrels out back for people to take. Sure enough, there are several bright blue, fifty-five-gallon barrels. I select two labeled "Apricot Flavoring," with an ingredients list that looks harmless. With the barrels strapped to the top of my car, I look ready to enter the Baja 500.

Why are my husband and I, professional people living in an affluent country in the year 2010, building a rainwater-collection system? In part, because we feel that this low-tech, mostly recycled, definitely jerry-rigged system is the right thing to do. In effect, it adds plasticity; a garden won't succeed if it does not have water. It is also smart; why run the well pump when gravity can do the work for free? How well household rain barrels will moderate the magnitude of change in precipitation is unanswerable. But the perception that it could, that we could adapt with a simple technology, even benefit from the excessive rain, is response itself.

Look around at the machines we rely on. This computer, embedded with circuits, metals, and hard plastic, is now a necessity in American culture, and this chair, cushioned with polyester foam and fabric and running on plastic wheels. The kettle I boil water in for tea, the steel stove with iron grates from which the heat to boil the water comes. All of it relies on resources extracted

with fossil fuels and from mines thousands of miles
from here. Ironically, the practice of catching rainwater exposes them to me. They are problems of design because they originated in a time different from the Age of
Warming. They were made with little care for the energy
that went into them or the waste that was left on the
cutting-room floor.

William McDonough and Michael Braungart, an
architect and chemist respectively, wrote a book called
Cradle to Cradle: Remaking the Way We Make Things,
in which they elucidate new principles of design for a
sustainable future. Here, "nutrients," the materials that
go into manufacturing (computers, chairs, stoves, fifty-
five-gallon drums, even) are held to a new set of guide-
lines. For McDonough and Braungart, all material
objects should eventually equal "food," hence the use of
the term *nutrients*.

In McDonough and Braungart's world, a chair is designed to come apart, after its useful life, that is. When
a chair is manufactured it is constructed in a way that
its polyester upholstery can become new upholstery (or,
if made from cotton, wool, or hemp, can become soil).
Its plastic arms and rolling casters should easily come
apart in order to be used again. All the materials that
go into the chair become either technical nutrients
that cycle in industry or biological nutrients that cycle
in the biosphere. In our quest for goods, McDonough
and Braungart ask that we think of "all the children of
all the species for all time." Clearly it is an ideal. But it
provides direction for our work and orients us to believe
in our persistence.

This is why the rain barrels resonate with me. Like
growing food and generating electricity from the sun,

collecting water is an essential aspect of our lives. On average my family uses twenty thousand gallons of water per year. That we could easily generate 20 percent of that—for nonpotable uses such as watering our gardens or flushing our toilets—by arranging old barrels beneath a roof line makes the decision to do so a matter of common sense. But beyond meeting these needs, collecting rainwater represents one response to the heap of problems we will have to solve as we adapt to a time of warming.

Such household- and citizen-based actions are percolating up. These types of innovative approaches serve as the incubators for ideas that influence policy well beyond the household. For instance, lack of capital is the major barrier for homeowners who want to install renewable energy. Recognizing this, a group of citizens petitioned the City of Berkeley, California, to support solar on their homes, and Property Assessed Clean Energy (PACE) was born. Under PACE, homeowners receive loans from the city to install photovoltaics and solar hot-water heaters. The homeowners pay the city back over twenty years as an assessment on their property-tax bill. This way the installation becomes part of the durable infrastructure that carries across owners. PACE has been so successful that its originator, Cisco DeVries, then an assistant to the mayor of Berkeley and now the CEO of Renewable Funding, has begun 240 other PACE-style programs across the country. States like Florida, Missouri, and Minnesota have changed their laws to support PACE, and Renewable Funding has successfully lobbied Congress to make the federal tax code more sympathetic to solar investments at the household scale. Citizen-based initiatives have always been a part of social movements, and the current movement to adapt to a warming climate is no exception.

~~~

Greensburg, Kansas, was devastated by a tornado in 2007. On May 4 at approximately 10:00 p.m., a tornado 1.7 miles wide with 210 mph winds demolished the town, leaving eleven people dead and 90 percent of the remaining fourteen hundred residents homeless. Founded in 1884 by westward pioneers, Greensburg originated as a farming town. Hand-dug wells, wind turbines that pumped water for cattle and fields, and hearty individuals living in sod huts characterized the town's early history.

Weather forecasters had been spinning the story all day. People sensed the heaviness in the air and a general unsettled feeling descended upon them. As sirens alerted the citizens to take cover, softball-sized hail began to fall. People raced to their closets or cellars for cover, like Aunt Em and Uncle Henry in *The Wizard of Oz*. For fifteen minutes, the sound of a freight train barreling on top of them terrorized the people of Greensburg. Then it suddenly stopped. People emerged into a jet-black night and the rubble of devastation. Roofs had been lifted off their nails and spun into the air. Trucks had flown across streets and onto flattened houses. Restaurant signs were twisted like pretzels. Even the town's arbor was felled, turning the landscape back to prairie grasses and a handful of well-rooted cottonwoods.

I can imagine what people saw, felt, and experienced.

Having lived in another tornado state, Wyoming, I remember watching from the deck of my parents' home
while the sky darkened to a sinister charcoal gray and
rope-shaped funnel clouds descended. I remember the
blare of sirens, civil defense announcing a coming tornado, and crawling into a closet beneath the stairs to
take cover. Tornadoes are spectacular and horrific,
and the winds, the railroading winds that accompany
them, nightmare winds that burst windows and bring
icy hail that smash what is left, these are part of the
horror too.

The Greensburg tornado was apocalyptic and intensely disorienting. In minutes the town lost any resemblance to what it had been. But Greensburg did not
disappear. When newscasters asked if they would rebuild, town members responded "Why not?," as if the
lesson of the tornado was, paradoxically, that while they
had lost everything, they felt fortunate.

Greensburg is transforming itself, adapting to not
only a postapocalyptic landscape but to the environmental crises that surround every town. Like citizens elsewhere, they are rethinking the design of the American
lifestyle. In particular, they are designing buildings that
speak to their unique commitment to the future. With
the help of wealthy patrons (including the actor Leo-
nardo DiCaprio), Greensburg is building schools, a city
hall, and an arts center with ecological design. Ten gigantic turbines laid out in a wind farm outside of town
provide enough power for four thousand households, far
more than a town of nine hundred people needs. Solar
photovoltaics, geothermal power (employing subsur-
face temperatures to heat above-ground spaces), green
roofs, energy efficiency, and reclaimed materials—brick

and wood from destroyed buildings—depict the new Greensburg.

What is ironic is that some of these same design principles were practiced by early pioneers. Wind power, recycled materials, water conservation, and even green roofs existed in nineteenth-century Kansas. Is Greensburg going back or moving forward? "When going back makes sense, you are going ahead," wrote the author and farmer Wendell Berry.

Al Letson, a radio journalist, produced a story on Greensburg in 2010 titled "To the Stars through Difficulty." When he interviewed citizens, many reported finding strength after the tornado that they didn't know they had. Some tried to leave Greensburg, to get out of Tornado Alley and find a place where trees or tall buildings would buffer them from what they had lived through. They were looking for a way to be absolutely secure. But many could not stay away. The momentum of rebirth in Greensburg was as motivating as the green elements of the rebirth itself. Yet Letson also detected grief, a sorrowfulness that you hear when people are mourning for what they have loved and lost, usually a person, but sometimes an element in the landscape, a building, a favorite gathering place, a grove of trees.

Part of our adaptation to the Age of Warming is dealing with this grief as well. There will be disasters like Greensburg in our future. There will be neighbors and family lost, trapped under the weight of the houses they built or felled by the structures they sought shelter in. Groves of trees will fall. Shorelines will be lost. Houses will be flooded. And some species of plant, insect, bird, mammal, and amphibian will not exist for children to see one hundred years from now. We will mourn for

them too. But as these losses accrue we will also adapt—
resolutely, creatively, automatically. Like the people of
Greensburg, our will to rebuild will be there. To paraphrase the philosopher William James, we inhabit the
world we attend to. That world, and our attention to it,
starts at home.

CHAPTER 7

Self-Reliance 2.0

I lower the laundry airer—a clothesline on ropes and pulleys—from the vaulted living room ceiling above our woodstove. Standing on a chair, I uncoil the rope from its metal boat-hitch screwed to an exposed wooden beam. The act of undoing a figure-eight knot calls to mind uncleating my family's wooden boat from its mooring. In my memory, my father is in the small boat with the tiller in one hand and the mainsheet in the other, waiting for me to push off for a summer's sail.

I no longer live on the Atlantic coast, and I rarely sail. But doing five loads of laundry in an energy-conscious home can require its own bit of navigation.

Untied and heavy with dry clothes, the "laundraire"— my family's affectionate pronunciation of the laundry airer—drops abruptly to the floor. It is an Amish contraption composed of simple materials: a lightweight metal bar, end braces, four cotton lines, and a length of rope to raise and lower it. The metal bar stabilizes the weight of pinned wet clothes so that I can hang two loads at a time and take advantage of the woodstove's heat, especially at night. The halyard-like rope runs through the pulleys; the elegant, simple physics permits my ten-year-old to haul the laundry up. Standing from the floor and pulling hand over hand, she is a young pirate hoisting the sails.

Transitioning to the laundraire was not difficult. My family and I were accustomed to using the propane dryer sparingly, aware that dryers are one of the largest energy consumers among household appliances. We hung our laundry outside in the warm months, and when the weather turned cold we used a wooden clotheshorse, the kind made of light pine that expands like an accordion to a dozen four-foot lengths. We placed the clotheshorse in front of the woodstove but kept bumping into it, placed as it was in front of two easy chairs and access to the first-floor bathroom. Sheets and towels were problematic too and ended up draped over doors or hanging from chairs. As my family grew, we needed a carbon-free solution to drying our clothes six months of the year. It arrived in *Lehman's Non-Electric Catalog,* a publication offering goods designed and crafted in the horse-and-buggy culture of the Amish.

As America's rural landscapes became electrified in the early twentieth century, as farms sprouted wooden windmills and as water turbines were laid in streams, the Amish chose to remain without. For them, electricity equated to accessing worldly temptations, ones that could erode family and church life. "Be not conformed to this world, but be transformed by the renewing of your mind that ye may prove what is that good and acceptable and perfect will of God" (Romans 12:2). This is the Old Testament quote that guides the Amish's resolve to disconnect from the world and keep technology at bay.

For many sects is has worked for more than two hundred years.

The Lehman's catalog offers all manner of nonelectric products and human-powered devices. Stove-top waffle irons, walk-behind garden tillers, crosscut saws, and wooden bowls, buckets, and spoons are a selection of what the company carries. Its catalog supports a low-carbon life, because most of the mechanical products do not require electricity. Lehman's has traditionally served the Amish's ideology of minimal technology, one shared by some Mennonite and Mormon families as well. But nowadays the Lehman's catalog serves people like me too, people who are experimenting with self-reliance and the desire, and at times compulsion, to become carbon-neutral.

I am wrestling with my use of fossil fuels. In my case, it is my car, propane generator, and cookstove. For others, the electric line that runs between their homes and coal, nuclear, and natural gas plants is their Achilles' heel. As Americans, we are inescapably caught in a culture of complicity; nearly everything we own and use has a carbon footprint. Backing out of this dilemma, limiting our consumption of hydrocarbons and transitioning to the end of fossil fuels, will be a historic cultural shift not unlike the way America ended its slave economy or the way workers broke with the tyranny of factories and unionized. It will happen by degrees, but it is sure to happen, because it carries such moral weight.

The Amish value of simplicity and self-reliance attracts me as I consider my chronic carbon complicity. Whether I choose low-carbon technologies or lifestyles that don't contribute carbon to the atmosphere at all (and even better if I find ways to take up carbon, like planting cover crops in my garden), I'm looking to minimize my impact while staying current in contemporary life—engaging in politics, listening to new music, and watching films, for example. I am not interested in becoming a hermit or dedicating all my waking hours to subsistence. Rather, I am choosing low-carbon and no-carbon technologies as a way forward, as part of a new definition of progress. At times this means forgoing comfort and convenience to stay oriented to my principles. Other times it means employing nineteenth-century technologies until twenty-first-century ones with similarly low impact come along. Hence the laundraire.

When the climate movement began two decades ago, many advocates anticipated an international accord that would limit carbon emissions worldwide. Activists and negotiators pointed to the Montreal Protocol, the international treaty that phased out ozone-depleting compounds in 1989. It took five years to write, was eventually ratified by 196 nations, and is the most successful act of international cooperation regarding the environment. This was the diplomatic precedent people in the movement meant to replicate.

Climate activists in the United States, however, have long speculated that in the absence of strong federal legislation, building grassroots support for carbon regulation across America—in communities, schools, churches, synagogues, and mosques—would be a necessary first step. Once the grassroots movement was established and a sympathetic president (and Congress) was in place, they could work swiftly, galvanizing change from the top down to meet what had been developed from the bottom up.

It has been two decades since the first UN Climate Change Conference (officially known as the International Conference of the Parties) met. But based on the outcome at Copenhagen in 2009, we are still worlds away from an international agreement, perhaps as far away as we have ever been. Our responsibility to control greenhouse gas emissions and our adaptation to their effects is increasingly a civic one, more personal and community-focused than regulatory. While it is naive to think that civic action alone will solve this problem—there is no doubt that it will take top-down policy changes, eventually—in the meantime, bottom-up actions are where change is taking place.

Conference goers gathered first in a chapel. Two-hundred-year-old white wooden pews with brown shellacked seats held the more than one thousand enthusiasts of self-reliance: Vermont organic farmers and gardeners. Knit caps and woolen vests, muddy boots, and rough hands were the outward signs that they made a living from the land, a rare type when fewer than 2 percent of Americans are farmers. But another, less identifiable, population of people was also present. Wearing standard khakis and fleece jackets, more L. L. Bean than Farmer's Supply, these were the other folks: people who garden small plots with relish or own backyard "farmlets." The group who in their spare time chase the goals of sufficiency and self-reliance.

The farmers' conference is held annually on a February weekend. It is a time of year when as an aspirant of self-reliance, I am anxious to plant again. My food stores are waning, the bottom of my freezer is in sight, and the firewood is two-thirds burned through. If the growing season the previous year precluded much social time, and winter is a time of lying low, then in February it feels good to be in community again and plan the next year's garden.

People pile in with canvas bags and warm clothing. Bundled babies nurse contentedly while their parents listen to speakers and chat with fellow farmers, gardeners, and homesteaders. Like other conventions of similarly minded people—the Rotarians, soccer enthusiasts, or modern-day dowsers—people unreservedly enjoy the fellowship of an orienting philosophy.

Per tradition, the conference director begins the day with a song: a ballad to agricultural life. A banjo player joins her, and as we stand I see people I recognize: a dairy farmer who keeps a small herd of Devon cows and produces a rich soft cheese I eat all winter; a vegetable farmer who specializes in squashes; another, farming cooperatively, who grows garlic, onions, potatoes, and leeks, and varieties that are "good keepers" and last all winter. There are meat producers who raise grass-fed cattle and lambs, pigs, and chickens, selling as much to local markets as to specialized butchers in New York City.

And then there are the orchardists growing acres of apples and pears. Many inherited their stock from others who weathered the rise of globalized food but sold out before the resurgence of eating locally. I chat with one grower who is diversifying into stone fruits: plums, cherries, apricots, and peaches. She's taking her chances with warmer-climate varieties. Like a goose that no longer migrates, she is banking on her recent experience of the seasons. Heeding the trend of warmer winters, earlier springs, and longer summers, she is adjusting her crops accordingly.

After the opening song, the audience disperses to workshops. Suburban Homesteading is my first choice, and it promises to achieve locavore goals with "Marco Polo realities," all those foods like pepper, pasta, and coffee that early explorers brought to the European palate. The presentation is made by a husband-and-wife team. In their midfifties, they hold full-time jobs and are as fiery about backyard sustainability as Pentecostalists reading from the good book. The couple raises chickens (for meat and eggs), honeybees (they bring in samples of sticky honeycomb, medicinal pollen, and beeswax candles), and garden beds teeming with vegetables (ones grown from seed in their 350-square-foot greenhouse, attached neatly to the garage).

The partners' enthusiasm is as contagious as their ideas are infinitely practical. They discuss how to sift homemade compost using the basket of an old grocery cart, and how PVC tubing fitted beneath a portable chicken-tractor prevents it from getting stuck in the muck. The packed audience finds it hard to contain their questions. "How much was the hoop house?" someone interrupts. "How deep did you dig the fence to keep out foxes?" another questions. "Eighteen feet," comes the response, and a groan goes up as pencils record the details and others calculate cost.

Everyone wants to share their backyard experiments and gather ideas for the coming season. The tenor of the discussion is equal parts extension service, 4-H club, and church revival. While the emphasis is on the practice of growing food, the discussion is framed as a collective response to the wide range of threatening phenomena of our times: the corporatization of food, peak oil, and global climate change among them.

Two generations ago, my mother's family were farmers. They made their livelihood growing potatoes on Prince Edward Island, Canada, a windswept island in the Atlantic Ocean north of Maine. My grandfather grew up on a near-treeless farm working alongside his six siblings, most of whom stayed on PEI to fish, farm, and, later in the twentieth century, cater to tourists.

My grandfather, Preston MacDougall, trained in the seminary but left the path to the priesthood to move "down under," to Boston, where he built bridges, ran a corner store, and owned a series of tidy aluminum-sided homes in towns that surround the city: Medford, Stoneham, and Somerville. As far as I can recall, he never gardened. The only remnant I remember from his island upbringing was a fiddle that hung above a plaid couch in his basement shop.

But like other young men and women in the early 1900s who immigrated to America from agrarian communities, Preston MacDougall was anxious to leave the farm behind. *Mechanization, progress,* and *production* were the words that spurred immigrants to build the industrial infrastructure of a still-young America. There were machines to man, bridges to build, and new cities to raise. Farming was a way of the past and seen as lesser work for a generation whose children would surely live better lives, own cars and houses, and move beyond the "backwards" nature of farming.

Ironically, as his granddaughter, living eight decades later, I am eager to grow my own food and experiment with farming. Barely two generations distant from Preston's clear departure from all things agrarian, and the self-reliance that came with it, I measure my own life by the gains in sufficiency that I can make. I teach my children about growing food. Together we plant seeds, tie up raspberries, and water fruit trees. We take garden walks where I instill in them an understanding of the cycles of the garden season. How the mizuna and spinach come first, how to pluck the early tomato flowers to increase plant height, how to squish the vermillion-colored eggs of the potato beetle, and how to plant beans with squash and corn. These are old agricultural lessons, taught in contemporary times.

But it is not all romanticism. Like others in rural, suburban, and urban settings, I am undertaking the work of producing food even while I recognize that my raised beds, small orchard, and clusters of berry bushes fall short of what my family consumes in a year. Still, it is a way to localize my response to the ecological constraints around me, even if their scale exceeds my imagination.

In this moment, I cannot slow Himalayan glaciers from melting, nor can I bring rain to the drought-stricken Central Valley of California. What I can do is become more self-reliant in my home place. Moreover, I can advance the same for my community by helping to create community gardens, supporting local food and energy cooperatives, and thinking systemically about the prudent use of resources for my small town. While I am deeply frustrated with my government's response to the climate crisis at hand, and have not abated my advocacy for change at larger scales, I find that more is accomplished by working at the local scale. It is also intensely satisfying.

"The perfect is the enemy of the good." I frequently run up against this aphorism when I assert the need for greater self-reliance. What counts? Do I make progress by driving to the farmers' market to buy from nearby

producers and then make an additional trip to the grocery store for paper goods? Is it better to cut down trees
shading the garden than to leave them standing to sequester carbon? And then there are the tough personal
questions, like how often do I visit my family (mother,
father, and two brothers) in Colorado, a long-haul plane
trip? The tentacles of a global economy based in fossil
fuels are long and far-reaching.

We cannot be perfect, but we can be honest. The
truth is it is impossible to be completely self-reliant and
live a contemporary life. But a perfect approach is, at
this time, not the goal. A transparent one is. "No man
is an island, entire of itself; every man is a piece of the
continent, a part of the main," wrote John Donne. What
I wish for is a way to ameliorate the crisis of our time
and head in the right direction, moving the trajectory toward actions that recognize ecological constraints, support living systems, and build grassroots momentum for
paradigmatic change.

Still, I am intrigued to observe the ways that self-
reliance emerges in American culture with the gains of
the twentieth century all around us. In the Age of Warming and under the threat of peak oil, people are experimenting with back-to-the-land lifestyles, raising bees in
their front yards and on the roofs of their apartments.
Others are building makeshift root cellars and buying
chest freezers to put up the abundance from their gardens and the copious extra that abounds in the fall.

By cultivating the essentials of our lives (or by knowing the person who cultivated them for you), we gain
a liberty from mass corporatization, and the line between production and consumption blurs. It feels good
to pick a head of lettuce, one grown without pesticides

or herbicides, and wash and eat the crunchy leaves. Being self-reliant is about identity and pleasure and, in a society that leads in the all-too-sedentary field of information technology, it stretches the muscles of work and imagination.

In effect, self-reliance is an adaptive strategy. In the same way a mosquito's biological clock helps it cope with warmer temperatures, self-reliance is a cultural trait that helps us cope with changing local conditions. The more we rely on our home places and communities to meet our essential needs of food and energy, the more readily we will recover from shock. The quest for self-reliance in the twenty-first century is about preparing for surprise and being resilient in the face of it: surprise hurricanes, surprise wildfires, surprise floods. After each, we will need to restore our equanimity. By pursuing self-reliant lifestyles we protect ourselves against those shocks and make a buffer where there was none before.

Our emerging experiments in self-reliance, everything from wood-heated laundry to electric cars run on home power, help propel what ecopsychologists call the "Great Turning," an ethical and practical transition that is based in environmental ethics. Furthermore, self-reliance helps redefine the relationship between humanity and nature. In contrast to controlling nature, seekers of self-reliance realize that the effect humans have on eco- and planetary systems is an evolutionary one. Humanity is coevolving with nature just as nature is evolving in response to human action. Human vulnerability is in the *perception* that we can control nature (with the added irony that while we instigate massive global change, we continue to expound control, pursuing, for example, geoengineering schemes like sending vast amounts of sulfur dioxide into space to reflect sunlight back into the atmosphere, or seeding the oceans with iron to bolster the production of carbon-absorbing phytoplankton).

~~~⌒⸏~~~

Every spring I take out from the library John Seymour's *The Complete Book of Self-Sufficiency*. First published in 1976, it is a handbook for urban, suburban (one acre), and rural (five acre) "smallholdings," an English term Seymour uses to describe a small plot of land. Colored pencil drawings illustrate self-sufficiency in each of these settings: a brick row house adjacent to community gardens; a cottage with fenced plots; a farmhouse surrounded by sheds for cows and chickens. The pictured straight-lined rectangular gardens bordered by hedges are characteristically English, and each chapter features a trade in self-reliance: raising rabbits, building cold frames, keeping bees, making cider, butchering a pig, drying herbs, tanning leather, shearing sheep, and thatching a roof, all the usual components of an agrarian life.

The book is as romantic as it is practical, and I imagine myself living in the perfectly ordered farmlet of Seymour's imagination. There I would step out my door to a world arranged in flawless squares of agro-ecological activity. Compost would be confined to a tidy corner, orchard trees would be espaliered against a southern wall, and young lettuces would soak up the humidity from glass cloches, cold frames that look like jewelry cases for vegetables. I pore over *The Complete Book of Self-Sufficiency*, reading it before bed the way my daughters pore over a large-format *Alice in Wonderland*. It is fantastical, ideal even, but there are elements that I'll dream into being, that propel me toward self-reliance as conditions change around me.

Since Seymour's publication, a second wave of self-sufficiency books has arrived to entice us to get close to soil, sun, and seed. Books such as *Sufficient: A Modern Guide to Sustainable Living*, by Tom Petherick, and *Self-Sufficient-ish Bible: An Ecoliving Guide for the 21st Century*, by Andy and Dave Hamilton, draw readers for their how-to approach. Like Seymour's, these books are reference materials, but they come with blog sites and shopping-cart menus where you can buy the tools for self-sufficiency.

When I search for books on self-sufficiency, another set of titles pops up. *Patriots: A Novel of Survival in the Coming Collapse*, by James Wesley Rawles, is one of them, as is *Emergency Food Storage and Survival Handbook: Everything You Need to Know to Keep Your Family Safe in a Crisis*, by Peggy Layton. These books contain much of the same information on butchering, canning, and beekeeping, but carry the tone of survival rather than the crunchy métier of self-sufficiency. If the former set of books envisions material wealth as measured by the weight of grapes pressed into an unlabeled bottle of homemade wine, the latter measures it in the months a stocked pantry could hold out.

This mix of offensive and defensive self-reliance is not new. The back-to-the-land movement of the 1950s and '60s (paved by Helen and Scott Nearing, who homesteaded in Vermont before moving to Maine) was infiltrated by bomb-shelter builders and nuclear war survivalists who imagined living through a postapocalyptic

America. The disparate groups shared practical wisdom but were driven, as they are today, by different ideologies and visions of the future.

If *self-reliance* can feel historic and descriptive of early America, then *sufficiency* is a little more accessible in present day. Self-reliance is personified in the way that pioneers—their fortunes tied to their wagons in seed corn, treadle sewing machines, and John Deere's early plows—set out from eastern towns for points west. These rugged individualists sought independence, in the near-perfect articulation of the word; there were few others to rely on.

Sufficiency, on the other hand, evokes the idea of having enough. Similar to *sustainable,* the term *sufficient* describes a person who is neither thriving nor failing but is consistently maintaining, whether it be with respect to personal economics or work. This is a clear departure from the dominant growth model of American culture, which expects a consistent increase year after year.

This revision of progress toward self-sufficiency, which includes self-reliance and emphasizes living comfortably but not extravagantly, is emerging from the sustainability movement. And one can see why. Americans are the most avid consumers in the world. In almost all sectors, from energy to water to material wealth, we are far and away the hungriest people. "Consumption no longer unifies a nation and elevates the individual, but, as practiced, undermines the biophysical foundations on which nations and individuals rest," writes the social scientist Thomas Princen in his book *The Logic of Sufficiency.* According to Princen, people are connecting the limits of the planet to personal behavior, limiting ourselves in the quotidian aspects of our lives. How much we travel, purchase electronic gadgets, and consume resource-intensive services are calculations people are making. The stunning awareness that both production and consumption of fossil fuels are responsible for the largest limitation the planet has thus far seen—the atmosphere's ability to store carbon—is upon us.

Finding ways to replace consumerism, the built-in reward for our working lives, is not without its difficulties. Yet people find that not consuming also has benefits. Time is number one. By consuming less we have less opportunity to make impulse purchases and arguably less exposure to the messages that encourage us to buy what we don't need. Less consumption means more time for leisure, for experiences in nature, for gardening, music, and the arts generally, activities that use nominal resources. More time to parent is also a benefit. I can attest to the breathing room that comes with a four-day workweek, a godsend for parents, as it increases the probability that someone is home for sick days, school concerts, and afternoon homework sessions.

Princen suggests that striving for sufficiency is a kind of self-management. In effect we are being asked to control our quest for more, an urge that seems quintessentially American. He provides evidence that when communities work collectively and provide for the good of many rather than wealth for a few, individuals restrain themselves. This occurs, for example, when fishermen, who manage a resource collectively, let that resource recover rather than become overexploited to the point of crashing.

There have been times in the past when as a nation we restrained consumption for the greater good. Thrift and frugality are universal qualities of the generation that lived through the Great Depression, people who were forced to restrict their consumption. I still hear stories of how central heating systems—radiators and all—were removed from the estates of wealthy Vermonters so that the metal could be contributed to the war effort in the 1940s. Collective restraint brings about social cohesion; think of how the Amish have remained a cohesive community by resisting the encroachments of modern technology.

These historical moments of restraint are not that different from the check on spending that occurred beginning in late 2007, when a recession affected millions of people who, because of a series of bad banking practices and unregulated financial schemes, lost their jobs and the value of their homes declined. In response, Americans consumed fewer goods, and for the first time in a half century carbon emissions fell 7 percent in 2009. As importantly, people rediscovered thrift and temperance and the appeal of working less. Four-day workweeks were offered as a way to cut budgets, and workers were asked to take unpaid leave. While this was not always conducive to individual feelings of well-being—people complained of reduced paychecks and upset schedules—many responded positively to the liberty of time and a new balance of life and work.

The idea of sufficiency is evolving. In an ecologically constrained world, the decision to work toward having "enough" wealth and curb the desire for more is becoming a rational response. But how do we determine what is enough? Enough, according to Princen, is contrary to insatiable consumption and also to extreme self-reliance. Enough is that middle ground where we conform to a sustainable rate of exploitation of resources held in common: a region's water supply, the ocean's ability to hold carbon, the world's biodiversity, and fertile soil. "Like peace, democracy, and progress, *sufficiency* is a big idea," writes Princen. A big idea that, coupled with self-reliance, manifests in a personal response to another big idea: how best to transition and persist through the greatest challenge humanity has ever faced, the Age of Warming.

———◦◦——

Today's self-reliance and sufficiency movements are not limited to rural America; "back to the land" is replaced with "back to the garden," including rooftop gardens, in urban landscapes. The proximity of people's living spaces, and the easy ability to see how others choose to live their lives, allows good ideas to spread quickly.

Take my friend Kate, an elementary school teacher in Baltimore who purchased a blighted house in 2009 during the mortgage crisis, when one in three Baltimore homeowners defaulted. The gutted structure had water in the basement, a leaky roof, no plumbing, and single-pane windows. But her goals to establish a self-reliant lifestyle that other city residents could emulate were clear, and first among them was that her four hundred square feet of land (not including containers on the porch) would produce the majority of her food. Second, the house's modest one thousand square feet, an area half the size of most new homes in America, would house her and two teenage children. She chose state-of-the-art energy-efficiency measures and a level of insulation that would keep her cool without air conditioning. Finally, she insisted that her heat would come from one cord of wood salvaged from fallen limbs and neighbors' pruning, material that would otherwise be destined for the landfill.

In essence, Kate built a prototype of urban self-reliance. And she did it on a public school teacher's salary. "Green design is not elite or for the elite, despite the impression you might get in *Architectural Digest*," Kate wrote me as Baltimore soared into 101 degrees Fahrenheit. "I want to show my neighbors and the community of bankers, developers, and city policy writers how to shrink their carbon footprint. Shrinking my own feels so good, it is the right thing to do given what I know, though I understand it isn't enough."

Kate had no idea how far-reaching her project would be. "My neighbor, Eugene, stopped his truck just to call out to me that his tomatoes in pots were doing fine and that he's already planning to till some earth for next year's crop. A neighbor's son comes home from college with an environmental studies degree and is ushered to my door for a tour. Several neighbors bring me their compost, and Da'Quan, a kindergartner from next door, plants his end-of-the-year Sugar Pie Pumpkin in a corner of my garden reserved for kids. He comes with his brother nearly every day to water his quickly burgeoning vine."

Kate's approach to self-reliance embodies how people in cities are adapting in place. On an urban block in gritty Baltimore, Kate confronted the same resource constraints that surround each of us, and she invented a different way to live—ecologically, philosophically, and socially—that others could mimic. People like Kate and her city neighbors are personifying the Great Turning and getting the transition under way. Moreover, they're speaking about it from the sidewalk, blogging about it on the Web, and designing for themselves what the new normal will be.

~~~

Human civilization has made historic transitions in the past. In the Stone Age, an era lasting approximately one hundred thousand years, humans lived in small villages and tribes, relied on hunting and gathering, and used language to communicate. Human transition into early civilization, an event anthropologists say occurred roughly eleven thousand years ago, brought with it the city-state organization of society. Agriculture replaced hunting and gathering and humans developed the ability to write. More recently, in the last one thousand years, much of the world transitioned into the modern era. Here, social organization was directed by the nation-state, rather than the city-state, and human economies became industrial. Printing revolutionized the way people communicated.

In a 2002 report, Paul Raskin and colleagues argue that we are now in the midst of a third significant transition, one they refer to as the planetary phase. Here, social organization transcends the nation-state and is directed by global governance. The dominant economy is a globalized one, and communication includes the Internet and electronic transmission of information. "Historical transitions are complex junctions, in which the entire cultural matrix and relationship of humanity to nature are transformed," these authors write. "The world system today overlays an emergent planetary dynamism onto modern, pre-modern, and even remnants of Stone Age culture."

Many people sense the transition to the planetary phase. The United Nations, the International Court, and the European Union, and treaties such as the Kyoto Climate Treaty and the Montreal Protocol, are testaments to global governance. Similarly, international trade agreements like NAFTA (the North American Free Trade Agreement) and CAFTA (the Central American Free Trade Agreement), and financial structures such as the World Bank and the International Monetary Fund, all support a globalized economy.

Yet changes are also occurring at the level of human consciousness and human conscience. Humans, as never before, are living across spatial scales. These scales begin with the personal sphere, the one over which we have the most influence. Here, self-reliance and experimentation with sufficiency are ours to execute. Change your showerhead, install energy-saving lightbulbs, eat within one hundred miles of your home, limit your use of inefficient appliances. All these actions occur in the personal spatial scale.

The next spatial scale up is the neighborhood or town. Here we export our personal influence and guide, and are guided by, our neighbors. Neighbors see the clothes-drying rack someone has set up in a living room, they taste the fresh lettuces grown in a cold frame in April, and talk together at the mailbox about the earliness of the spring. At the bus stop, I ask my neighbor how he controls powdery mildew on his grapes. He asks me how much the solar panels I installed on my roof cost and the mileage my car gets. Neighbors govern a landscape together even while privately owning the pieces. As we transition to a new era, the neighborhood becomes a collective area of influence.

Beyond the neighborhood and small town are the foodshed and watershed. Our spatial awareness of nearby lands and waterways helps us understand how we affect our local environment. When I apply phosphorous to my lawn (or to the powdery mildew attacking my grapes), I contribute to algal growth in the pond downstream. When I purchase energy from the wind farm one county over, I support the production of electricity that doesn't pollute the soil with nuclear waste or toxify lakes with mercury.

Like the neighborhood and town, one's foodshed/watershed is large enough that we can see the effect of our collective actions. It also allows us to ask: am I encouraging ecological health in the place bounded by water and productive land? If the locavore restrains himself to eating foods within a one-hundred-mile radius of his home, then the area of influence is thirty-one thousand square miles. This is the next spatial scale.

Despite our technological prowess with mapping and navigation systems, like the Global Positioning System device that impersonates a dictatorial female (Australian, at that) telling you when to turn, the spatial scales beyond home, neighborhood/town, and foodshed/watershed begin to enter the realm of abstraction. It is difficult for our consciousness to spatially comprehend larger and larger geographic regions. The bioregion, for instance, is the next spatial scale that we exist in. It is a series of repeating ecosystems that characterize a region. Where I live, the deciduous forests of the Northeast describe my bioregion.

Our cultural and ecological identity is tied to our

bioregion. From this vantage point we see how our collective actions matter to the region's health and its ecological persistence. Working together to save unique natural communities within the bioregion—a bog with rare orchids or a prairie pothole that serves as a refuge for waterfowl—these actions align us with the greater landscape and the unique qualities of bioregions beyond ours.

Each of these scales—personal, neighborhood/small town, watershed and bioregion—demands civic engagement. The last two spatial spheres are the hemisphere and globe. I add "hemisphere" because as a global population we are aware of how different the sources of emissions between the hemispheres are. By and large, humans living in the northern hemisphere have contributed the lion's share of carbon. In contrast, humans living in the countries that are still developing in the southern hemisphere, with the exception of China and India, are responsible for less than 10 percent.

Finally we reach the global scale. Few of us believe that the actions we take have a measurable effect on the functioning of the planet. With almost seven billion people in the world, it is the great accumulation of actions—personal, industrial, governmental—and their effect over time that result in the phenomenon of global ecological crisis. Yet it is the consciousness of each of us as members of Earth's biotic community that compels people to act and reverse these trends.

There have been moments in our environmental history when we understood the boundarylessness of pollution. For me one of the most powerful was when I was a nursing mother. During that time I learned that breast milk, if regulated like infant formula, would violate US

Food and Drug Administration standards and could not be sold as food. Women's breast milk had become so laden with persistent pollutants that it could be deemed a toxic substance. As if this were not damning enough, women living in the Arctic had higher levels of toxicity in their breast milk than any other human population. Yet many of these women ate a traditional diet of seal, whale, and fish. Pollutants, it turns out, become concentrated in the fats of these organisms, and toxins biomagnify up the food chain. Thus their traditional diet works against them. But the question remained: why would industrial pollution be concentrated in the Arctic? The mystery was solved when researchers discovered that wind currents move pollutants northward, in a process called global fractionation. They get caught up in evaporation and condensation cycles, migrating north until cold temperatures restrict their movement into the hydrologic cycle and they remain fixed in the Arctic's biota.

There are other examples of the boundarylessness of global pollution. Twenty-eight years post-Chernobyl and fifteen hundred miles from the nuclear reactor, lambs grown on 382 British farms have levels of radioactive cesium that prohibit them from being sold. And miles of plastic are trapped in an enormous ocean whirlpool called the North Pacific Gyre, a collection of plastic waste estimated to be twice the size of Texas, which threatens seabirds, sea mammals, and the smallest of marine organisms that ingest or get trapped in the plague of debris.

But few pollutants rival the boundarylessness of carbon dioxide. Here an odorless, invisible gas travels globally. Circulating in our atmosphere, it goes everywhere and is long-lived. What we once thought was a benign molecule, an essential plant nutrient, the stuff of our own exhalation, now endangers anything alive.

Our carbon complicity, expressed in the language of our conscience, motivates us to change our behaviors, enacting cultural adaptations that effectively deal with our concerns and create new ways of being. Hanging the laundry above the woodstove, planting fruit trees, greening an urban dwelling, and installing photovoltaics on a school's roof are all expressions of social obligation, a social obligation that translates across spatial scales.

There is no doubt that actions based in self-reliance and in ideas about sufficiency will not alone change the world. Certainly not during the lifetimes of the people living today. Yet they make measurable differences in our lives, in our identity, and in the values we develop in others. The world is transitioning. If feminism saw the personal as political, then now the personal is also paradigmatic. The Great Turning is ours to enact.

CHAPTER 8

The Pragmatism
of Adaptation

Helen, my daughter, is experimenting with closed systems. She received a terrarium for her birthday, a two-gallon glass cylinder with a heavy lid, the kind that usually comes with a sharp metal scoop and multicolored jelly beans to sell by the pound. She and I have layered the base of the terrarium with shards of charcoal, stones from the road, and an inch of soil. In the woods, we have found spongy club moss and a maidenhair fern with spiraling jet-black stems. These we have planted in the column of glass. To this she is about to add a torn piece of bark from a paper birch. It is the color of her skin, with holes the size of small beads where beetles have drilled their escape. She holds it up to the sky like a mini telescope and sunlight shines through.

In her terrarium, Helen has placed a nautilus shell called "snake's eye." It sits up against the glass, evoking a time when fern, moss, and mollusk lived together and the shore of an inland sea approached our woods. Finally, she adds a three-inch-long red eft, the immature stage of the eastern newt. "What will the newt eat?" she asks me as she lowers its gummy, bendable body into the terrarium, already steamy with transpiration.

Helen's experience with nature, the cool moisture of the newt's skin, the way its flame-orange spots lie across its back, and the feeling she has of holding a wild creature—notions of power, escape, rareness—all add to the experiment. She is beginning to understand what is wild, what is tame, and what her role is in caring for another being in this world. These concepts are subtle, but still they are taking hold in her five-year-old self. I know when she asks me, "How long can I keep him?" The red eft has already become her domain—part pet, part progeny.

I explain that wild things need more space than a terrarium affords, more than a beetle to eat, and more than a clump of moss for habitat. "But I'm protecting it," she insists, speaking defensively yet pragmatically. This is her first notion of stewardship; she can be the overseer, can guide the diminutive being into adulthood. Yes, I think, your impulse is correct, you can take care of it, just not in this way.

⟞⟋⟍

Pragmatism emerged after the Civil War as a philosophical as well as a practical tool against which truth and belief could be tested. "When we are happy with a decision it doesn't feel arbitrary; it feels like the decision we had to reach; it fits with the whole of social understanding, self understanding, and moral weight. We know we're right and then we know why," wrote the pragmatist John Dewey.

The originators of American pragmatism were William James, a philosopher and teacher at Harvard University, and Oliver Wendell Holmes, a Civil War veteran and US Supreme Court justice. These men, in addition to the more reclusive but intellectually productive Charles Peirce, formed an approach to values and ideas that encouraged people to experiment and experience the world to derive their own set of beliefs. If Ralph Waldo Emerson, a predecessor to the pragmatists, emphasized the importance of a person's conscience in the confrontation of social issues, then the pragmatists felt that personal experience was how one developed a belief system.

While this approach to philosophy was informed by the political arguments and deep moral divide that followed the Civil War, it was also informed by the publication of Charles Darwin's *On the Origin of Species*. In his seminal book, Darwin argued for the theory of natural selection, an idea that challenged the dominant worldview that species were placed on Earth by a great creator. The pragmatists argued that decisions, like species responding to natural selection, *fit* the social and moral world of a person's life, very much the way the presence of a pollinator *fits* with the flowering of a tree or the way a squirrel's birthing schedule *fits* with warm temperatures and cone abundance.

In the twentieth century pragmatism gained prominence as a range of scholars and social activists were guided by its principles. John Dewey, a scholar and author of books on democracy and education, brought American pragmatism to his laboratories at the University of Chicago, where he experimented with how children learned. It was in Chicago that Dewey met Jane Addams, a social worker and founder of Hull House, a settlement for recent immigrants and the poor. Addams's working theory was that to understand the experience of the poor one needed to live it alongside them. This was the beginning of the concept of "care ethics," empathetic moral action that connected people—the philanthropist, social worker, and person needing care—around a social issue. Working together, Addams and Dewey challenged one another to relate pragmatism to the civic responsibility of being American.

Seventy years later, America was developing an environmental ethic. Deep ecology, biophilia, ecocentrism, and animal liberation spawned ethical conversations for a movement that began in 1965 with Rachel Carson's *Silent Spring*. The rise of environmental ethics challenged the deeply humanistic perspective of civic pragmatism, encouraging people to expand their empathy to the natural world and the diverse ecosystems it contains. Like the care ethics shepherded by Jane Addams, environmental ethics grew to encompass the entire circle of life and raised the elemental issue of existence.

By the end of the twentieth century, environmental ethics had ushered in a new line of thinking: environmental pragmatism. Like James's idea that pragmatism (though he preferred the term *practicalism*) was a way to interpret ideas and assess their truth for their practical consequences, environmental pragmatism was constructed as a methodology, an open-ended inquiry into the very practical problems humanity was wreaking on the environment. As Ben Minteer, the author of *The Landscape of Reform: Civic Pragmatism and Environmental Thought in America,* writes: "The most salient feature of pragmatism is its instrumental character and the emphasis it places on the realm of practice ... it takes shape as individuals—and communities—confront problems, learn about their (and others') values and beliefs, and adjust and progressively improve their natural and built environment."

In a world inundated with information, it is hard to hone in on one's singular experience of truth. Yet for people experimenting with the messiness of transition— what feels like fixing a bicycle while riding it at the same time—we are reporting a truth: mitigating and adapting to climate change is the right action. As Dewey noted one hundred years earlier, these actions *fit* our scientific and social understanding of the world, and we experience not only the truth of adapting to climate change but the adherence to a new set of values that are evolving as a result.

In 2005 Glenn Albrecht, an Australian psychologist, derived a new term for the experience people have when their home places no longer feel like their own. This term is *solastalgia*. When there is environmental disruption and a community like the Canadian Inuit experience the ice melting and their way of life ending, they experience solastalgia. When subsistence farmers in Ghana can no longer predict the planting season because of radical changes in rainfall, they mourn for what their lands used to be like. They experience solastalgia.

Like nostalgia, solastalgia has elements of homesickness, sentimental yearning, and the memory of another time. It comes from the Latin root words *solatium,* meaning comfort, and *algos,* meaning pain. But it is distinguished by the fact that the people describing the emotion have never left home. Rather, the experience of degradation has unfolded in front of their eyes.

Albrecht's naming of this psychological condition

has elicited acknowledgment of it in people all over the world, who are experiencing solastalgia in their own landscapes. We can see, touch, smell, and feel the planetary changes. We realize that the scale of change is different from what we've experienced before. It is all around us, inescapable, a trajectory with momentum that we'll need to adapt to.

While there are no glaciers where I live, variable and extreme weather is having an effect on my landscape. While single storms cannot be causally proven to be due to climate change, the weather is decidedly in line with the predictions: wetter, warmer, and more variable. Take the spring of 2010. Historic warmth in March brought 80 degree Fahrenheit temperatures (26 degrees Celsius), but by the end of the month, record snows (twenty-two inches in my yard) and freezing temperatures set in for nearly a week. A local apple orchard lost 60 percent of its fruit when temperatures dipped below 25 degrees Fahrenheit (-4 degrees Celsius) and forest trees in the band of elevation around eighteen hundred feet lost their young leaves to freezing. Two months later a dark band, similar to what I see after a first frost in November, stretched from eighteen hundred to twenty-two hundred feet. The contrast with the month of May's consummate green was unsettling. What was a band of death doing in the midst of the season when life unfurls?

Ecopsychologists wonder how deeply our minds and the health of our psyches are affected by the current scale of planetary change. And it is not only local events, like mountaintop mining in Appalachia or clear-cutting in Oregon, but global events that make us anxious for the health of the planet. As empathetic persons we are losing sleep over ocean acidification, the extinction of amphibians, and desertification in China, to name a few.

In the small town where I live, there is a single elementary school, a blue vinyl-sided one-story building that sits on the edge of a hay meadow. An American flag hangs out front and the older children raise and lower it in the morning and afternoon. A playground is spread out in a side yard where swings and a hard plastic jungle gym have been installed.

The cafeteria/gym is the heart of the school. Children meet there in the morning, pile their jackets and backpacks, and collect in groups to build with Legos, draw with crayons, or eat a warm cinnamon roll, sold on Fridays to raise funds for the school lunch program. Life-sized snowmen and colored-paper trees whose leaves display the four seasons decorate the painted, cement-block walls.

The school, like many others across America, serves and establishes community. When the roof leaks, local carpenters donate their time to make the repairs until a bond measure can be passed and a new roof bought. When the front gardens need tending, a group of parents gather to pull crabgrass and lay mulch, all the while catching up on local politics and the administrative style of the new principal. But the community is fiscally conservative, and any additional projects, including the new playground, must come from outside funds.

One day a third-grade teacher asked me if her class could visit my house. They were studying electricity and wanted to see a solar home. A couple of weeks later, four carloads of third graders, parent chaperones, and teachers sat cross-legged in my living room. They pushed the toaster, played a CD, turned on a lamp, and read out the watt hours consumed for each appliance. They stood around the batteries in the basement and asked where they came from. They looked up at the laundraire and found it "cool."

Soon after, I became determined to normalize this experience for the children at our small school. The visit had made me appreciate how novel solar energy still was for kids. Seeing it in action, in a home with a tree house and garage much like their own, was enough for the children to understand the simple reality: solar works. It could work in their houses too, as well as in the offices, barns, and computer plants where their parents worked. It could even work at their school.

I gathered three parents to help me raise money to put in a solar system at the school. We stipulated that it be visible to any visitor and be at the edge of the playground, where the children would see it every day. We dedicated ourselves to the mundane tasks of raising money (selling lightbulbs at the local hardware store) and writing grants to local businesses (the gas station, an ice-cream factory, a computer corporation), and to nonprofits that might consider funding such an effort.

In two years we had raised $20,000 and installed two solar arrays at the edge of the playground. The local newspaper took pictures of the children grinning as they crouched beneath the panels, sunlight reflecting off the sail-like rectangles of glass and silicon behind them. Working with the third- and fourth-grade teachers, a university student devised a curriculum, with topics including how solar energy works, the basic physics of electron transfer, and how power from the children's playground worked to power their school. A second group of parents started a school garden with the arrays situated in the center. Soon sunflowers were growing on either side, and raised beds of beets, carrots, basil, and tomatoes greeted the students when they returned in the fall.

When the parents first met with their vision for a solar array in the schoolyard, they hoped to bring renewable energy to the 150 children who attend the school. From the four-year-old preschoolers to the ten-year-old fourth graders, children in our small town would recognize solar energy and be familiar with how it works. Moreover, they would have experience that it works. Solar would be a part of the school the same way whole-wheat cinnamon rolls on Friday were. It would become the new normal.

The amount of electricity produced by the 4 kilowatts of solar at our school is minimal compared to the amount it consumes in a year; less than 5 percent. But it is the children's experience with the technology as a utility, with the sunflowers growing alongside it mimicking the system's ability to capture sunlight, that is important to them as students.

In fifth grade the children in our community are bused to a middle school ten miles away. For four years they travel back and forth before they attend a regional high school an additional ten miles away. In 2011 the middle school, with a population of five hundred children from four towns, will install 200 kilowatts of solar energy on its flat roof. With money from federal and state renewable-energy funds, the array will be one of the largest in Vermont.

Impressed by what our little elementary school had done, a well-connected group of individuals moved to the next level of change, increasing the resilience of our community with distributed renewable energy, advancing and normalizing the alternatives to fossil fuels, and pragmatically leading the transition as adults aware of the crises at hand and the need to adapt to them.

If self-reliance is a pragmatic step toward reconciling one's environmental concerns with a belief in the persistence of life, then establishing local economies in the midst of globalization is the way a new political economy could grow. Localism is, as writer David Hess states, "an alternative global economy that is more effective at addressing global problems of sustainability and justice than a global economy dominated by large multinational corporations."

People are searching for security as the world becomes a less certain place and as the agitation of transition gets under way. Local economies, like gardens out back, are a first step toward remaking our societies to be less about industrial consumerism and more about environmental sustainability, justice, and persistence.

Further, local economies support many of the practical considerations of self-reliance, and they provide a way forward for people frustrated by the plodding pace of federal regulation. Thus, establishing businesses that operate from the same set of principles that an individual holds can actively counter our sense of chronic complicity.

My neighbor Dean's bakery, for instance, lines up with my values. He offers a morning cup of fair trade coffee in a reusable mug; you can take it home if you like, just bring it back. Milk from a nearby dairy lightens my coffee and Dean's baked goods are made from local ingredients: eggs from chickens and ducks out back, flour from a Vermont-based wheat company, and a variety of fruits in season: rhubarb from a neighbor, plums from an orchard thirty miles away, and apples from his own tree, bent but still producing decades after being planted. Dean places my goods in a take-out bag. It looks and feels like cellophane but is made from cellulose—a plant carbohydrate—and will dissolve innocuously when exposed to water and sunlight (read: thirty days in my compost pile).

But it is the scale as much as the ingredients that deepens my satisfaction with Dean's bakery. Like my patronage of the farmers' market, the general store, the library, and the physician with an office in town, consumption at Dean's feels like a relationship of mutual support. There is nothing faceless or demeaning about it. Being a customer there benefits not only Dean but the people he supports in turn and the principles each of those entities agree upon. This is where the fruits of localization are realized and translated into progress for civil society well beyond the individual.

"A practical idealist," writes the author and Buddhist scholar Ken Jones, "is one who accepts her fear without being possessed by it. Living beyond optimism and pessimism, she is a patient and clear-sighted possibilist." I am a practical idealist, even though it is challenging to remain so in these times. The signals of ecological constraint are all around us, yet individual and civic responses to environmental degradation are also emerging. I recognize them and I am a part of them. I predict that in time they will coalesce into a social and adaptive response to climate change. This will be a genetic-like change that will spread through us to our offspring and become fixed in our culture.

Think for a moment of ideas that initially were met with resistance but later became mundane. In the early days of electrification, people recoiled at the idea of a lightning-like current (AC or DC) in the walls of their homes. When automobiles were first introduced, people assumed they would never travel more than fifteen miles per hour; otherwise they would scare the horses! But education and implementation eventually led to acceptance of technologies that are now commonplace. In the same way, practices that may have once gone unquestioned—the use of aerosol spray cans containing hydrofluorocarbon, littering—have become anathema, jarring our ethical sensibilities. Now the same   is happening with where and what we purchase.

The term *localization* describes the mission to raise environmental and social justice goals in the way we consume. Localists, having examined the potential to ecologically modernize the present state of global commerce, to "green" it as it were, see an inherent contradiction: almost by necessity, ecological limits constrain profits and will always be secondary to shareholders' interest. Localists conclude that developing alternative local economies is a superior approach. Small-scale economies, they argue, can be linked to one another in a kind of global localism (like buying fair trade coffee at my CSA pick-up) more effectively, more rapidly, and more ecologically than greening the dominant economic paradigm.

Much of the localization rhetoric expresses support for locally owned businesses; saving mom-and-pop stores is something localists advocate for. Yet if locally owned businesses are selling the same goods as the multinational big-box stores on the outskirts of town, there is marginal benefit to supporting them. Rather, the goal is to develop organizing principles for small enterprise that address the environmental and social ills that have emerged with global commerce.

But others see the world differently. They see apocalypse, the end times, and dystopia. James Lovelock, the author, most recently, of *The Revenge of Gaia,* writes, "The evidence coming in from the watchers of the world brings news of an imminent shift in our climate towards one that can easily be described as Hell: so hot, so deadly that only a handful of teeming billions now alive will survive."

Apocalyptic thinking is part and parcel of the human psyche. Now, public reference to an apocalypse triggered by global warming is more frequent. Extremely apprehensive language is rationalized if not outright believed. This ideology of fatalism is prevalent in the climate change rhetoric. It grows out of an urgency naturally imbued in climate science reports. Here, radical tipping points and runaway events are spoken about not only as foreseeable but as catastrophic.

The fatalists justify their urgency as necessary to spur action. Terms such as *posthuman* are used to describe the ultimate: the extinction of humankind. While I am sympathetic to the need for urgency, articulating the concept of climate apocalypse can solidify a reality that it was only meant to describe, notes the sociologist Eileen Crist. Furthermore, it may perversely reinforce and legitimize the theoretical concept, breeding an ideology of fatalism.

The fact is that climate science is entering unknown territory. Anticipating Earth's climate over the next one hundred years is daunting, even to climatologists. To understand the effect of anthropocentric climate change on Earth's systems, we will create new methodologies, new computers, and perhaps new complex sciences. In the meantime, we juggle the projections against our determination to find a way to adapt, evolve, and persist. We heed the urgency while resisting fatalism.

———⁂———

A mourning cloak butterfly flits along with me as I walk in the leafless forest on a spring day. How easy it is to be in the woods, both the butterfly and I, in these weeks with little growing but the woodland flowers. The dry leaves beneath us ruffle as we pass a dogtooth violet sheltered at the base of a birch tree. The butterfly looks for nectar and finds something in a wild gooseberry bush.

It is moments like these, as I walk with a butterfly on a spring day or take the kindergarten class to the stream to find caddis fly larva cases—grainy cylinders of sand glued to the bottom of rocks—that I look out at the world and am hopeful. I am hopeful that the future will be guided into a time of more persistence than extinction, more pleasure than suffering, more adaptation than despair.

———⁂———

The first industrial revolution began in the eighteenth century, when Britain's economy transitioned from horsepower to refined coal as the energy of mechanization. The discovery of modern petroleum (literally "rock oil") in the mid-1800s triggered a second industrial revolution. At first petroleum was refined into a type of kerosene, used primarily to light lamps. Later, in the early 1900s when the internal combustion engine was commercialized, petroleum fueled all manner of mechanization, from early farm equipment to factories, automobiles, and eventually airplanes.

Now fossil fuels, including coal, petroleum, and natural gas, form the basis of modern society. At any moment in time millions, perhaps billions, of pistons are firing. Untold combustion engines are running, and unfathomable sheets of plastic, rolls of polyester, and miles of asphalt are being processed, manufactured, and transformed into products. Our reliance on fossil fuels penetrates everything about our society. "There are virtually no extant forms of transportation, beyond shoe leather and bicycles, that are not based on oil, and even our shoes are now often made of oil. Food production is very energy-intensive, clothes and furniture and most pharmaceuticals are made from and with petroleum, and most jobs would cease to exist without petroleum," write Charles Hall and John Day in their article on the limits to growth after peak oil.

We realize that to continue our use of fossil fuels is morally wrong. Like the economies (and consumers) dependent on slave labor, we can no longer rely on the cheap but perilous use of petroleum, natural gas, and coal. New energies must be discovered, and the use of viable, carbon-free technologies must replace fossil-fuel-powered ones. This is indisputable.

All my life I have known that species go extinct due to human activity. As a young girl, I knew this when my father took me to see the sage grouse on the Laramie Plain outside of Cheyenne, Wyoming. There, as dawn graced the eastern edge of the high plains, we spotted the male grouse. Its tail feathers were arrayed in a perfect fan of earthy hues as it pranced about the hens, forming leks— groups of females—around him. They were attracted to his prowess, a signal of his fitness. But we knew that the bird was at risk. Cattle ranching, a practice that dominated the High Plains since they were settled by Europeans, had degraded and fragmented the landscape. Now sage grouse occupy only half of the habitat they did 150 years ago.

Species extinction is an epidemic. In combination with habitat destruction and invasive species, climate change is hastening Earth's sixth mass extinction. In response, the taxonomist and author Edward O. Wilson has begun an online *Encyclopedia of Life* to chronicle and record the existence, and sometimes the simultaneous departure, of species. Alan Pounds, a tropical ecologist who recorded the extinction of the golden tree frog due to climate change, is paraphrased by Crist when she writes, "Climate change is the bullet that threatens extinction but industrial-consumer society is pulling the trigger." Climate change is a symptom of overexpansion of the human species. It is an indicator of our blasé relationship to the ecological limits to growth. We need systemic change.

One evening I was asked to speak at a religious sanctuary where people from all denominations gather. On the blond maple walls were niches holding religious iconography: a Jewish menorah, a Christian cross, a Hopi dream-catcher, and the crescent moon and star, symbols of Muslim faith. Behind the podium was a wall of glass framing Lake Champlain and the Adirondack Mountains in the afternoon light. Farm fields and hay fields filled the space between sanctuary and water's edge, forming a lush carpet of fresh green growth. Two children played chase in the grass and the audience watched them as they listened to the speakers.

It was a small crowd, barely a dozen, but it included a handful of young people, several middle-aged adults, and two women in their eighties who sat up front. I was part of a panel that included a fine artist whose abstract paintings explore the human-nature relationship, and a dramatist dedicated to establishing eco-teams—support groups for ecological living.

The fine artist spoke eloquently about her paintings, the dramatist boomed with humor, and I read my writing. As artists, we relayed what we were thinking and feeling about the age of climate change. The audience listened kindly to the presentation, and then the question-and-answer session began. People were eager to talk, to tell their own stories of how they were confronting their complicity, how their actions reflected the urgency of the scientists, and how they were steering—from their own homes and with their children beside them—a different future.

"I lived through the Depression," said Ruth, one of the elders. "We lived with scarcity. We had to conserve because we didn't have much. How do we get others, my own family included, to behave differently?" A man in the back spoke up. He said he was a father with an eight-year-old and a six-year-old. He had joined an eco-team through his church. They egged each other on to limit their carbon emissions, checked up on whether they had followed through (changed lightbulbs, shortened the length of showers, found people to carpool with, insulated their homes), and formed a loose community that struggled together through the first phase of their adaptation. He said, "I've learned so much, especially how much else I can do. What I've done is not enough. I want to keep going."

Others spoke about solving the problem politically. They asked: Who would lead the transition away from fossil fuels? When will America fully embrace renewables and put a price on carbon? Does the rhetoric of climate crisis change human behavior or, as the father of two testified, does human behavior change behavior? Each expressed his or her acceptance that we were in the midst of a phenomenon that was no longer abstract; they saw global warming in their everyday lives. Whether it was a thunderstorm in winter or a deluge that flooded their basements in summer, people were unanimous in recognizing that the world was different. Climate change was no longer an event for the next generation to live through; they felt uniquely challenged to do right by themselves, their children, and their grandchildren. The imperative of social obligation, a motivation that was articulated through their conscience, compelled them. It was as if the room was filled with conscientious objectors who were against a war for the simple reason that it was wrong.

In 1777 the Commonwealth of Vermont—at the time a sovereign state—abolished slavery. Its Constitution, under the section "A Declaration of the Rights of the Inhabitants of the State of Vermont," read, "That all men are born equally free and independent, and have certain natural, inherent and unalienable rights ... Therefore, no male person, born in this country, or brought from over sea, ought to be holden [*sic*] by law, to serve any person, as a servant, slave or apprentice." When Vermont became a state in 1791, the abolition of slavery remained, preceding a national prohibition by seventy-two years.

Cultural transitions take time, because core values

are deep-seated. Cultures and subcultures are differentiated by their values, and values are hard to change. Whether human nature is good or bad; whether humans should be dominant, in harmony with, or subordinate to nature; and what sense of time we function in—learning from the past, being fully in the present, or planning for the future—are all values that affect our willingness to adapt. The importance of mitigating emissions does not necessarily resonate with a person who believes that an omnipotent deity will intervene to resolve our dilemmas. On the other hand, a person who believes we must be good stewards of God's creation responds positively to mitigation.

Dominant values in the United States will evolve to be consonant with climate change as well as the broad family of ecological constraints to which it belongs. I see this happening; the small but growing new culture seeks a profound shift in values. It seeks a path that breaks with consumption as an orienting modality and leads toward a prudent use of resources that meets our needs while preserving life on Earth.

This shift is instigated by the global condition humanity finds itself in. While experienced differently depending on where one lives, the condition of change is shared by all. Whether it is in the Arctic, where the native people have no name for the robins they see in the spring, or in New England, where spring comes three weeks earlier, we are struggling as a human community to chart our fate.

Charles Darwin influenced early American pragmatists with his science of adaptation. His understanding of how organisms adjust to dynamic conditions, and his view that change is continual, not directional, encouraged the pragmatists to think about human beliefs and truths in a similar light. Like the advances in fitness that occur as species adapt to changing conditions—the way the beaks of Galapagos finches evolve depending on the size of seeds in their diet, which in turn rests on the amount of rainfall the seed plants receive—we too are influenced by changing conditions.

Today we restrict our understanding of these moments of fitness to tight, elegant relationships in nature: the way a fawn's dappled back looks like sunlight on a forest floor and, because she gives off no scent, eludes large-nosed canines. Or how exact the length of a yucca moth's tongue is given the depth of the night-blooming yucca it pollinates.

But adaptation and fitness should not be limited to the nonhuman world. Like the pragmatists' notion that ideas adjust to fit the conditions at hand—the social, moral, and political world of one's time—ideas originating in our culture emerge to suit the conditions we are experiencing. And not only the ecological conditions of our home places but the planet as a whole.

Our response to a time of warming is the essence of pragmatism, in its most practical and philosophical sense. We act practically because the actions themselves reinforce new actions. Turning off lights leads to installing compact fluorescent bulbs which in turn leads to buying renewable energy which in turn leads to installing solar photovoltaics. Practical actions provide the basis for a new set of ideas and beliefs to live by. Like the eco-team father whose carbon-reducing behavior encouraged him to keep reducing, the collection of these behaviors reinforces the truth of the phenomenon that he believes in.

Citizens the world over recognize the profound eco-logical, social, and economic problems we face. How poverty describes one in four, how two billion people are without electricity, how one in three live on less than two dollars a day, and how we lose approximately seventy-four species each day to extinction. While the solutions to these problems are tremendously complex, the are-nas people have power over are where and how they live: their lifestyles, their homes, what they choose to eat and drink, and how they get their energy. Increasingly these actions are in the context of community, too. It is in this realm that the practice of mitigating and adapting to cli-mate change leaves the sphere of the ideal and enters the practical.

In the same way that human identity is shaped by our parents and peers, a new identity is emerging that is in sync with sustainability. This new identity diminishes the role of materialism and elevates the values of social justice and environmental health. It draws from our his-tory of self-reliance, the contemporary localization and transition movements, and the phenomenon of ecologi-cal constraints to establish new social norms.

This new normal is a salve to the chronic complic-ity that weighs on us. It demands that we find a fresh approach, a middle ground, where we live equitable, healthy lives while preserving space for nonhuman life to maintain its own. "Life wants to live. Life so com-pletely wants to live," writes the author Derrick Jensen, ". . . and to the degree that we consider ourselves among and allied with the living, our task is clear: to help life live."

It is in this direction that we become allies with the natural world, distinctly aware that our persistence depends on it. Perhaps climate change forces us to see this relationship unreservedly for the first time. Or perhaps we are developing a consciousness that reflects the scale of change that is under way. In either case, human adaptation to the Age of Warming will genuinely advance when we define ourselves by this alliance and without reservation help life live.

Persistence

I have no recollection of learning to walk. But when I saw my children learn, pulling themselves up on the corner of a table or balancing with one hand on the rung of a chair, I thought about how, from the age of one onward, we rely so entirely on our feet.

Here we are not alone. In fact, we are much like other animals who rely on their feet, and their toes too. All birds, for instance, have two feet, but their toes differ. Perching birds have four toes, three that face forward and one that faces backward, allowing them to clasp their feet securely to a branch while their bodies slump forward in sleep.

In contrast, birds like nuthatches and woodpeckers have two toes that face forward and two that face backward, an arrangement that stabilizes them when they climb vertically up trees. (A stiff tail helps too.) Ostriches, on the other hand, have only two toes and both face forward. One of the toes has even become a gnarly, hoof-like nail that presumably gains them trac-

tion as they race away from their lion predators at
30 mph.

Obviously, bird feet and toes are not all alike. A woodpecker does not perch and an ostrich does not climb
vertically up trees. Across the geologic time frames that
birds have existed—from the warm Cretaceous to the
glacial Pleistocene—their feet and toes have adapted to
accommodate a vast array of different environments and
the ways in which conditions have changed over time.

My brother lost the use of his legs, feet, and toes ten
years ago. As a person with multiple sclerosis, a neuro-
logical illness that attacks the myelin sheath on the spi-
nal column and interrupts messages from the brain to
one's limbs, Larry's body underwent a chronic and consistent decline. First he could not run. Then he could not
walk the stairs to his San Francisco apartment. Spastic-
ity, loss of bladder function, and diminished motor skills
in his hands and fingers followed. One day after accepting the wheelchair, he told me, "Looks like it's time to
hang up walking."

Without an able body, Larry adapted by living the life
of the mind. He took up new languages and read copi-
ously. His voice, eyes, sense of smell and hearing, and,
perhaps most importantly, his mental acuity are sharper
and more sensitive than before. What few of us feel we
could ever adapt to, Larry has. Yet like others who have
experienced this kind of profound, personal difficulty, he
is sanguine. "I would not trade what I have learned from
this illness. Not even the illness itself," he has confided.

But the transitions have not been easy. I remember
them as an unrelenting series he was forced to adopt,
each one slightly robbing him of his integrity even while
it bestowed new liberties: my grandmother's cane, an

accessory that smoothed his gait, made him older than his thirty-five years; clumsy metal cuffs and sticks that provided support could never be easily propped and they fell noisily to the floor in restaurants and movie theaters. And then there was his first wheelchair. Jet-black with red inner rims and athletic-looking tires, it stumbled over uncut curbs and teetered on ramps, threatening to tip over.

~ ~ ~

Climate change is one of multiple conditions that reflect the ecological constraints we are coming up against. Rather than resorting to fatalism and calculating the last mating pair of humans on the planet, the positivist in me looks for ways to adapt to these conditions and adopt new ways of being. This approach is founded on a belief in persistence and an ability to endure tremendous loss, to struggle through the horrors of disaster, and to support the suffering of those left behind.

I do not wish to sound naive. I fully recognize how urgent these times are, how vastly we have changed Earth's systems, and how complicated the resolutions to these problems will be. I also understand how limited the natural, physical, and social sciences are in predicting what the future looks like given this complexity, and further, that the dynamics of climate change, or the predictions of what is to come, are not as well understood as we would like them to be.

But our persistence is a brave one. It begins, as I've described in the preceding pages, by examining adaptation in life around us. What is our "drought escape" strategy? What is the human analog to resetting our biological clock when the growing season extends into the fall? How flexible can we be when resources change as suddenly as they did for the pelagic pelican? In the biological world around us, there is an unfathomable, and as yet untapped, resource for probing how to adapt to change.

Human adaptation will be integral to the social transition that will accompany us in the Age of Warming. Like social movements in the past—women's right to vote, the eight-hour workday, and civil rights for African Americans—the climate change movement and its goal to end carbon emissions will involve the reformation of our lives. I see this transition as having begun with the way we live, day to day. Growing our own food and purchasing it from people who use ecologically sound practices personalizes and then embodies the transition before us. How we use energy and the creative ways we are weaning ourselves from oil, gas, and coal are next. Orienting ourselves to the sun's near-infinite supply of energy, and finding ways to power our lives with it, brings a clear-eyed confrontation to the discussion of where our energy comes from.

We can frame our lives this way: we are adapting to a warming climate. At moments, nineteenth-century practices like the laundry airer, bicycle, and push mower are our best options. At other moments, we embrace rain barrels and cold frames, clumsy, perhaps, but utilitarian. In time, we will add twenty-first-century technologies that are fully informed by the ecological and social predicaments we are in: a hydrogen fuel cell run on a solar array that powers our homes and cars, bullet trains that replace short-haul flights, and buildings that are so well designed and integrated into their environments that they take little to no new input of energy to heat or cool. And all along, I see the philosophies and practices of self-reliance, sufficiency, and pragmatism supporting the belief that the transition relies on individual action as a reflection of our commitment to stay.

Higher ground is ours to find. For some it is upwards in elevation or polewards in latitude, even requiring migrating to another country altogether. For others, gaining higher ground will come by means of a raised garden bed or a set of water barrels below a downspout. Higher ground will be found in grafting new varieties of grapes onto native stock or raising rice in submerged fields. It will be attained by generating solar power, driving electric vehicles, and installing green roofs. By bicycling to work or, when possible, staying put and working from home.

But higher ground also lies in territory beyond these pragmatic actions. It is in our determination to care about what we love, to protect life that is threatened, to grieve for what is lost, and to believe we can endure the Age of Warming. The biological and cultural environments that we have depended on in the past will undoubtedly change. But the adaptations we bring into existence will be the very makings of our persistence.

Acknowledgments

For conversations about climate change, adaptation, and persistence, from ecological and cultural perspectives, I thank Ben Falk, Gary Meffe, Sandra Steingraber, Walter Poleman, Bill McKibben, John Elder, Jon Isham, David Ward, Deirdre Heekin, Bill Shutkin, James Kuntsler, Gail Boyajian, Tom Stearns, Lawrence E. Seidl, and Daniel Goodyear.

I wish to thank the Vermont Studio Center for a writing fellowship in the spring of 2010. It was a fruitful time and came when I needed it most. I also thank my fellow writers at the Center for their comradeship and for discussing the manuscript with me. Ann Armbrecht provided advice and support on the writing process throughout the book and I thank her for that.

I sincerely thank my editor, Alexis Rizzuto, for her deep interest in all things ecological and for her passion for the subject of climate change; I look forward to the projects ahead of us. I also thank my agent, Russell Galen, for supporting this book from the beginning and for putting it in the larger context of our work together as he so ably does. In addition, I thank the Beacon staff for their continued support of my writing and for the publication of beautiful and timely books.

My sincere thanks to Ann Seidl for her creative and meticulous edit of the manuscript. I also thank the rest of the Seidl family for their encouragement, honest comments, and unambiguous but gentle criticism.

This book would not have been written if it were not for my family and the way in which our lives benefit one another's. From the bottom of my heart I thank my husband Daniel Goodyear for being the chief engineer of our home experiments in adaptation and for giving me time to put words to paper. In this book I was able to describe the world as I see it and, as importantly, as we live it.

Notes

CHAPTER 1: ADAPTING TO A CARBONATED WORLD

The term *Anthropocene* was coined by Paul Crutzen and Eugene Stoermer, in "The Anthropocene," *Global Change Newsletter* 41 (2000), to describe an epoch in which the global effect of human activity was likened to geologic events. The term has subsequently been used by Jan Zalasiewicz et al. in "Are We Now Living in the Anthropocene?" *GSA Today* 18, no. 2 (2008).

James Hansen and colleagues wrote the article "Target Atmospheric CO_2: Where Should Humanity Aim?," *Open Atmospheric Science Journal* 2 (2008).

Realizing the perilousness of increasing carbon dioxide concentrations, Bill McKibben, an author and climate activist, founded the organization 350.org, a group dedicated to policies that use 350 ppm as the universal goal of carbon dioxide reduction even while acknowledging that the world has overshot it. By the time the United Nations Conference of the Parties met to negotiate a new international climate treaty in December 2009, carbon dioxide concentrations in the atmosphere were 390 ppm.

There is an established tension between adaptation and mitigation. David Orr, a conservation biologist and environmental scientist who advocates for the direct application of ecological science to society, writes, "Adaptation must be a second priority to effective and rapid mitigation that contains the scale and scope of climatic destabilization." Orr argues that adaptation takes away from mitigative efforts; for him, support for adaptation encourages technical optimism and experimentation with dangerous geoengineering schemes. Embracing adaptation with the kind of gusto implicit in these schemes could, according to Orr's thinking, derail us from solving the root of the problem: reducing carbon emissions. To read more of Orr's thinking on adaptation and mitigation, see "Baggage: The Case for Climate Mitigation," *Conservation Biology* 23, no. 4 (2009). Roger Pielke et al., "Lifting the Taboo on Adaptation," *Nature* 445 (2007), offers an opposing argument.

See Remy de la Mauviniere and Elaine Ganley, "Natural and Human Factors Made French Storm a Killer," Associated Press, March 4, 2010. The reference to the speed of mitigation versus adaptation comes from the United Kingdom's Carbon Disclosure Project (2009). Rio de Janeiro's extreme weather and David Zee's interpretation of it is discussed in Fabiana Frayssinet, "Environment—Brazil: A Tragedy of Local and Global Dimensions," Inter Press Service (IPS), April 9, 2010.

See Tim McClanahan, Nicholas Polunin, and Terry Done, "Ecological States and the Resilience of Coral Reefs," *Conservation Ecology* 6, no. 2 (2002), for more on coral reefs and climate change.

For information on the underestimation of climate change events see the article "Adaptation" in *Climate Change 101: Understanding and Responding to Global Climate Change,* published by the Pew Center on Global Climate Change and the Pew Center on the States series (2008).

Nicholas Stern wrote *The Economics of Climate Change: The Stern Review* (Cambridge, UK: Cambridge University Press, 2007).

My thinking on adaptation, vulnerability, and resilience was informed by Gina Ziervogel et al., "Adapting to Climate Variability: Pumpkins, People and Policy," *Natural Resource Forum* 36 (2006).

Al Gore's *Earth in the Balance* (New York: Houghton Mifflin, 1992) was one of the first books to popularize the issue of global warming and press for carbon mitigation while reserving adaptation as a way forward.

Susan Solomon and colleagues published their article "Irreversible Climate Change Due to Carbon Emissions," in *Proceedings of the National Academy of Sciences* 106, no. 6 (2009).

For more information on Inuit and Norse response to the Little Ice Age in Greenland, see Jared Diamond's *Collapse: How Societies Choose to Fail or Succeed* (New York: Penguin Books, 2005); Brian Fagan's *The Little Ice Age* (New York: Basic Books, 2000); L. K. Barlow et al.'s "Interdisciplinary Investigations of the End of the Norse Western Settlement in Greenland," *Holocene* 7, no. 4 (1997); and Andrew Dugmore et al.'s "Norse Greenland Settlement: Reflections on Climate Change, Trade, and the Contrasting Fates of Human Settlements in the North Atlantic Islands," *Arctic Anthropology* 44, no. 1 (2007).

See the USDA's 2000 planting zone map as compared to the Arbor Foundation's 2009 revised planting map. By comparing the two maps one can see that many areas in the United States have changed zones for planting. In Vermont, many areas now rated as zone 4 and 5 were rated zone 3 and 4 only fifteen years ago.

My thinking on human vulnerability and the idea of resilience was informed by Stephané Hallegatte, "Strategies to Adapt to an Uncertain Climate Change," *Global Environmental Change* 19 (2009).

CHAPTER 2: FITTING IN

For a complete explanation of Arthur Weis's experiments on *Brassica rapa* (mizuna), see Steven Franks, Arthur Weis, and Sheina Sims, "Rapid Evolution of Flowering Time by an Annual Plant in Response to a Climate Fluctuation," *Proceedings of the National Academy of Sciences* 104, no. 4 (2007).

Carl Zimmer's "First Comes Global Warming, Then an Evolutionary Explosion," *Yale Environment* 360 (August 3, 2009), http:// www.e360.yale.edu/content/feature.msp?id=2178, speaks briefly about the evolutionary changes in red squirrels and mizuna, among other species.

For a survey of the effects of climate change across species, I consulted Camille Parmesan and Gary Yohe's "A Globally Coherent Fingerprint of Climate Change Impacts across Natural Systems," *Nature* 421 (2003). Further, in my book *Early Spring: An Ecologist and Her Children Wake to a Warming World* (Boston: Beacon Press, 2009), I recount phenological differences occurring throughout America's landscapes, with a particular focus on New England.

For an understanding of how species will adjust and/or adapt to climate change, I consulted Reed Noss, "Beyond Kyoto: Forest Management in a Time of Rapid Climate Change," *Conservation Biology* 15, no. 3 (2001).

For further discussion of phenotypic plasticity and evolution, see Trevor Price et al., "The Role of Phenotypic Plasticity in Driving Genetic Evolution," *Proceedings of the Biological Sciences* 270 (2003). In addition, Mary Jane West-Eberhard's *Developmental Plasticity and Evolution* (Oxford, UK: Oxford University Press, 2003) is an excellent volume on evolution and our contemporary understanding of plasticity and genetic change.

All temperature increases reported here come from *Climate Change 2007*, a report of the Intergovernmental Panel on Climate Change (IPCC), unless otherwise noted.

Fortunately, some organisms are adapting to the degree of climate change we've seen thus far and have the genetic variation on which natural selection can act. Unfortunately, others are less able to adapt, less able to fit in to the variable and novel environments around them. These organisms are becoming locally extinct. In 2004 Chris Thomas and colleagues published a paper predicting that between 15 and 37 percent of the world's species would go extinct by 2050 because of climate change. The results of their climate modeling, in which they analyzed the reactions to anticipated climate-change events of over one thousand plants and animals from different regions, alarmed ecologists. How could ecosystems be expected to function with this level of loss? Yet subsequent analysis reached different conclusions. When computer models employed a finer spatial scale, a 25-by-25-meter area of interest versus a 16-by-16-kilometer area, species' persistence was more likely than extinction. Moreover, when researchers look closely at the landscape, even as conditions change, they see diverse topographies and microclimates that aren't captured when the lens is too wide. In one case a coarse scale model predicted complete loss of habitat for plants, but when the scale was refined, the outcome was 100 percent persistence. The two papers I relied on here were Chris Thomas et al., "Extinction Risk from Climate Change," *Nature* 427 (2004), and Kathy Willis and Shonil Bhagwat, "Biodiversity and Climate Change," *Science* 326, no. 5954 (2009).

My information on red squirrels as model organisms for the study of phenotypic plasticity, microevolution, and climate change comes from Stan Boutin's webpage at the University of Alberta, http://www .biology.ualberta.ca/faculty/stan_boutin/?Page=734, and from Boutin's paper, with coauthors Denis Réale, Andrew G. McAdam, and Dominique Berteaux, "Genetic and Plastic Responses of a Northern Mammal to Climate Change," *Proceedings of the Royal Society of London* B 270 (2003).

Ernst Mayr's *The Growth of Biological Thought: Diversity, Evolution, and Inheritance* (Cambridge, MA: Harvard University Press, 1982) was essential to my understanding of the history of evolution. Written nearly thirty years ago, it remains a classic piece on the history of biological sciences. In particular, I drew from Mayr's chapter "Evolution before Darwin," including his citation of G. Rousseau, "Lamarck et Darwin," *Bulletin du Museum National d'Histoire Naturelle* 41 (1969).

Cassandra Brooke discusses the definition of adaptation in the context of climate change on both biological and cultural realms in "Conservation and Adaptation to Climate Change," *Conservation Biology* 22, no. 6 (2008).

Wendy Van Doorslaer and colleagues write about temperature as the dominant factor affecting species and how it serves as a selective force in algae in their paper "Adaptive Microevolutionary Responses to Simulated Global Warming in *Simocephalus vetulus:* A Mesocosm Study," *Global Change Biology* 13, no. 4 (2007).

I was first introduced to Goethe's saying in Stephen Jay Gould's book *Full House: The Spread of Excellence from Plato to Darwin* (New York: Three Rivers Press, 1996) and Gould's discussion of it in the book's last chapter on evolution in human culture.

Botanical descriptions of pitcher plants came from the website of the International Carnivorous Plant Society, http://www.carnivorous plants.org/, accessed on September 22, 2009.

My understanding of pitcher plant mosquitoes and the plants they inhabit comes entirely from the research of William Bradshaw and Christina Holzapfel and their articles, including "Genetic Response to Rapid Climate Change: It's Seasonal Timing That Matters," *Molecular Ecology* 17, no. 1 (2008); "Tantalizing Timeless," *Science* 316, no. 5833 (2007); "Evolutionary Response to Rapid Climate Change," *Science* 312, no. 5779 (2006); and the article, with colleagues Kevin Emerson and Sabrina Drake, "Evolution of Photoperiodic Time Measurement Is Independent of the Circadian Clock in the Pitcher-Plant Mosquito, *Wyeomyia smithii,*" *Journal of Comparative Physiology A* 195, no. 4 (2009).

The idea that culture can be a selective agent is also discussed by Deborah Rogers and Paul Ehrlich in "Natural Selection and Cultural Rates of Change," *Proceedings of the National Academy of Sciences* 105, no. 9 (2008).

Kevin Laland, John Odling-Smee, and Sean Myles discussed how culture affects human evolution in "How Culture Shaped the Human Genome: Bringing Genetics and the Human Sciences Together," *Nature Reviews Genetics* 11 (2010). I also drew from Nicholas Wade's reporting of the subject in "The Twists and Turns of History and of DNA," and "Human Culture an Evolutionary Force," *New York Times,* March 12, 2006, and March 1, 2010, respectively.

Recent adaptations in the human genome were understood by reading Benjamin Voight et al.'s "A Map of Recent Positive Selection in the Human Genome," *PLoS Biology* 4 (2006), www.plosbiology .org.

The quote from Alain originated in *Propos d'un Normand, 1906– 1914* (Paris: Gallimard, 1952) and is translated in Rogers and Ehrlich (cited above).

CHAPTER 3: ON MIGRATION

My thoughts on migration were grounded in readings from Robert Ricklefs's text *Ecology* (New York: Freeman, 1990); R. McLeman and B. Smit, "Migration: An Adaptation to Climate Change," *Climatic Change* 76, nos. 1–2 (2006); and E. Mexe-Hausken's "On the (Im-)possibilities of Defining Human Climate Thresholds," *Climatic Change* 89, nos. 3–4 (2008).

I consulted Daryl Wheeler and Paul Ehrlich's *The Birder's Handbook: A Field Guide to the Natural History of North American Birds* (New York: Simon and Schuster, 1988) in my research on pelicans. In addition, I referred to the article "The Biogeography of California Brown Pelicans (*Pelecanus occidentalist californicus*)" by Elise Willett, San Francisco State University, Department of Geography, November 28, 2001, http://bss.sfsu.edu/holzman/courses/Fall01%20 projects/BPELICAN1.HTM.

David Ward's "Change in Abundance of Pacific Black Brant Wintering in Alaska: Evidence of a Climate Warming Effect?," *Arctic* 62, no. 3 (2009), added greatly to my understanding of the current situation for these migratory birds.

The predicted number of climate refugees was relayed in a review article by Carolina Fritz titled "Climate Change and Migration: Sorting Through Complex Issues without the Hype," published online at the Migration Policy Institute's Migration Information Source, March 2010, http://www.migrationinformation.org/Feature/display .cfm?ID=773.

See William H. Frey, Audrey Singer, and David Park, *Resettling New Orleans: The First Full Picture from the Census* (Washington, DC: Brookings Institution Metropolitan Policy Forum, September 2007), http://www.brookings.edu/~/media/Files/rc/reports/2007/07katrina freysinger/ 20070912_katrinafreysinger.pdf.

Sander E. van der Leeuw's "Climate and Society: Lessons from the Past 10,000 Years" and Eric Blinman's "2000 Years of Cultural Adaptation to Climate Change in the Southwestern United States," both in *AMBIO: A Journal of the Human Environment,* special report no. 14 (2008), informed my thinking on past adaptations to climate change.

My understanding of the level rise as predicted by the IPCC and contrasted with new science was based on Walter Pfeffer et al.'s "Kinematic Constraints on Glacier Contributions to 21st-Century Sea-Level Rise," *Science* 321, no. 5894 (2008).

See C. X. Li et al., "Development of the Volga Delta in Response to Caspian Sea-Level Fluctuation during Last 100 Years," *Journal of Coastal Research* 20, no. 2 (2004).

"Chain migration" is discussed in Sabine L. Perch-Nielsen, Michèle B. Bättig, and Dieter Imboden, "Exploring the Link between Climate Change and Migration," *Climatic Change* 91, nos. 3-4 (2008).

See Stephen Willis et al., "Assisted Colonization in a Changing Climate: A Test-Study Using Two UK Butterflies," *Conservation Letters* 2 (2009).

The Jeremy Rifkin quote is from his *The Empathetic Civilization: A Race to Global Consciousness in a World in Crisis* (New York: Penguin Books, 2010).

CHAPTER 4: FEAST OR FAMINE

For more information on Ben Falk's permaculture design, see his company's website, www.wholesystemsdesign.com.

There are many definitions of *adaptation* circulating, but the definition I use here for adaptation in the context of agricultural production comes from S. Mark Howden et al., "Adapting Agriculture to Climate Change," *Proceedings of the National Academy of Sciences* 104, no. 50 (2007).

My discussion of droughts was greatly informed by Giorgios Kallis's review of droughts as a socioenvironmental phenomenon in "Droughts," *Annual Review of Environment and Resources* 33 (November 2008).

Linda and Takeshi Akaogi live in Westminster West, Vermont, and farm ten acres on their Akaogi farm. They raise vegetables, honey, chickens, and rice organically and sell to local markets. Their US Department of Agriculture grants, grant proposals, and rice-growing manual can be obtained by calling Northeast Sustainable Agriculture Research and Education at 802-656-0471.

See Aldo Leopold's *A Sand County Almanac* (Oxford, UK: Oxford University Press, 1949).

A description of Monsanto's climate-ready genes for corn was taken from the company's website on November 17, 2009, https://www.monsanto.com/responsibility/our_pledge/healthier_environment/carbon_sequestration.asp, under the heading "Monsanto can help agriculture keep carbon in balance."

For more on agricultural technology as a panacea in a warming world see Giorgios Kallis's review of droughts, cited above. For more information on thresholds and adaptive response I used Mark Howden et al.'s "Adapting Agriculture to Climate Change," *Proceedings of the National Academy of Sciences* 104, no. 50 (2007). I also consulted Wolfram Schlenker and Michael J. Roberts, "Nonlinear Temperature Effects Indicate Severe Damage to US Crop Yields under Climate Change," *Proceedings of the National Academy of Sciences* 106, no. 37 (2009).

Cary Fowler writes about thresholds and crop diversity in her article "Crop Diversity: Neolithic Foundations of Agriculture's Future Adaptation to Climate Change," *AMBIO: A Journal of the Human Environment,* special report no. 14 (2008).

My thinking on foodsheds was expanded by reading Christian Peters et al., "Foodshed Analysis and Its Relevance to Sustainability," *Renewable Agriculture and Food Systems* 24, no. 1 (2008). Further, my information on the history of community-supported agriculture is from Katherine L. Adam, "Community Supported Agriculture" (2006), available at National Sustainable Agriculture Information Service, http://www.attra.org/attra-pub/csa.html.

I first spoke about living "true to life" and deep pragmatism in a keynote address delivered at the 2006 Winter Conference of the Northeast Organic Farming Association of Vermont. Subsequently, I wrote about these ideas in the article "True-to-Life" for *Vermont Commons* 14 (June 2006).

Results of the Rodale Institute's research on conventional versus organic, and till versus no-till agriculture, are in Lotter et al.'s "The Performance of Organic and Conventional Cropping Systems in an Extreme Climate Year," *American Journal of Alternative Agriculture* 18, no. 3 (2003).

For a thorough overview of how climate change, as reported in the IPCC's 2007 report, will affect agriculture worldwide, I used Mannava Sivukumar and Robert Stefanski's "Climate Change's Adaptation, Mitigation, and Sustainability in Agriculture" and Raymond Motha's "Developing an Adaptation Strategy for Sustainable Agriculture," both published in the journal of the Hungarian Meteorological Service *Időjárás* 113, nos. 1–2 (2009).

CHAPTER 5: OUR OLDEST AND NEWEST ENERGY

Craig Whitlock, "Cloudy Germany: A Powerhouse in Solar Energy," *Washington Post,* May 5, 2007; and Caroline Bayley, "Germany's Sunny Revolution," BBC News, January 10, 2008, www.news.bbc.co.uk, describe the state of the solar industry in Germany.

US Solar Market Trends is published by the Interstate Renewable Energy Council. These annual reports include per capita solar installation in the United States and can be accessed at www.irecusa.com. Peter Fairley, "Ultra-Efficient Photovoltaics," *Technology Review,* June 15, 2007, describes how efficiency gains will come from using new materials to make photovoltaic panels.

Data on coal production and consumption in the United States is from the Energy Information Administration, an office of the US Department of Energy, www.eia.gov.

Margaret Palmer et al. present a consensus to end mountain-top removal and valley filling to access coal streams in the southeast United States in "Mountaintop Mining Consequences," *Science* 327, no. 5962 (2010).

For more information on estimates of natural gas reserves, see Energy Information Administration, "World Proved Reserves of Oil and Natural Gas, Most Recent Estimates," http://www.eia.doe.gov/emeu/international/reserves.html.

The 2008 US Department of Energy report on wind generation is *20% Wind Energy by 2030: Increasing Wind Energy's Contribution to US Electricity Supply,* www1.eere.energy.gov/windandhydro/pdfs/41869.pdf.

John Farrell and David Morris authored *Energy Self-Reliant States: Homegrown Renewable Power* in 2009, with an updated version in 2010: www.newrules.org/energy/node/2526.

Daniel Nocera is the Henry Dreyfus Professor of Energy at the Massachusetts Institute of Technology. A description of his research and current projects can be found at www.mit.edu. In addition, I used his paper "Chemistry of Personalized Solar Energy," *Inorganic Chemistry* 48, no. 21 (2009), as well as Kevin Bullis's interview with Nocera in *Technology Review,* May 2007.

For more information on Google's solar-panel arrays and plug-in vehicle research, see www.google.org.

CHAPTER 6: LOCALIZING HOME

To learn more about the Living Building Challenge, consult the website of the recently formed International Living Building Institute, at www.ilbi.org.

The idea of cycling industrial wastes into nutrients to be used again and again is taken from William McDonough and Michael Braungart's *Cradle to Cradle: Remaking the Way We Make Things* (New York: North Point Press, 2002).

I used Don Alstad's "Population Structure and the Conundrum of Local Adaptation," in Susan Mopper and Sharon Y. Strauss's edited volume *Genetic Structure and Local Adaptation in Natural Insect Populations: Effects of Ecology, Life History, and Behavior* (New York: Chapman and Hall, 1998) as the example of local adaptation in ecological systems.

The quote from Bill Reed comes from a keynote address he delivered at the national convention of the American Institute of Architects, San Antonio, TX, May 2007.

John Gendall's "Architecture That Imitates Life," *Harvard Magazine,* September–October 2009, informed my understanding of HOK's project in Lavasa, India.

On sustainable design, see McDonough and Braungart's *Cradle to Cradle,* cited above.

"The Incubators," written by Douglas Fischer for the blog *Daily Climate,* May 20, 2010, http://wwwp.dailyclimate.org/tdc-newsroom/2010/05/the-incubators, relays how citizen-based actions are influencing policy, including PACE.

The quote from Wendell Berry comes from his *Farming: A Hand Book* (New York: Harcourt, Brace, Jovanovich, 1970).

Al Letson's radio show is *State of the Re:Union,* archived at http://www.prx.org/series/30960-state-of-the-re-union.

CHAPTER 7: SELF-RELIANCE 2.0

My thinking on resilience and its relationship to self-reliance was largely informed by Carle Folke et al., "Resilience and Sustainable Development: Building Adaptive Capacity in a World of Transformations," *AMBIO: A Journal of the Human Environment* 31, no. 5 (2002).

The John Donne quote is from Meditation 17 of his *Devotions Upon Emergent Occasions* (1624).

Thomas Princen's *The Logic of Sufficiency* (Cambridge, MA: MIT Press, 2005) greatly informed my understanding of self-sufficiency as a principle that contemporary notions of sustainability may be founded on. Discussion of the benefits of not consuming was informed by Joe Dominguez and Vicki Robin's *Your Money or Your Life: Transforming Your Relationship with Money and Achieving Financial Independence* (New York: Penguin, 1992).

Paul Raskin et al. write of the planetary phase in *Great Transition: The Promise and Lure of the Times Ahead,* a report of the Global Scenario Group (Boston: Stockholm Environment Institute, 2002).

The discussion of systems thinking was drawn from James Kay and Eric Schneider, "Embracing Complexity: The Challenge of the Ecosystem Approach," *Alternatives* 20, no. 3 (1994); and Donella H. Meadows, "Places to Intervene in a System (In Increasing Order of Effectiveness)," *Whole Earth* 91 (Winter 1997).

Walter Rogan wrote *Epidemiology of Environmental Chemical Contaminants in Breastmilk* (New York: Plenum Press, 1986). Information on toxicity levels of breast milk in the Arctic came from Marla Cone's *Silent Snow: The Slow Poisoning of the Arctic* (New York: Grove/Atlantic Publishing, 2006).

Information on British farms and the Chernobyl nuclear disaster came from Helen Caldicott's book *Why Nuclear Power Is Not the Answer* (New York: New Press, 2006).

CHAPTER 8: THE PRAGMATISM OF ADAPTATION

My understanding of pragmatism was deeply informed by Louis Menand's historical account of the early pragmatists in his book *The Metaphysical Club* (New York: Farrar, Straus and Giroux, 2001), from which the John Dewey quote was taken. As a newcomer to the world of environmental ethics, I relied heavily on Robyn Eckersley's "Environmental Pragmatism, Ecocentrism, and Deliberative Democracy," in Ben A. Minteer and Bob Pepperman Taylor, eds., *Democracy and the Claims of Nature: Critical Perspectives for a New Century,* (Lanham, MD: Rowan and Littlefield, 2002); Patrick Curry's *Ecological Ethics: An Introduction* (Cambridge, UK: Polity Press, 2001); and Ben Minteer's *The Landscape of Reform: Civic Pragmatism and Environmental Thought in America* (Cambridge, MA: MIT Press, 2006).

The field of ecopsychology is contributing to our understanding of the effect of climate change on people's mental health. Articles by Daniel B. Smith, "Is There an Ecological Unconscious?" *New York Times Magazine,* January 27, 2010; and Glenn Albrecht, "Solastalgia: A New Concept in Health and Identity," *Philosophy, Activism, Nature* 3 (2005) and "Solastalgia: Distress Caused by Environmental Change," *Australasian Psychiatry* 15, no. 1 (2007), contributed to my thinking.

David J. Hess wrote *Localist Movements in a Global Economy: Sustainability, Justice, and Urban Development in the United States* (Cambridge, MA: MIT Press, 2009), a book that describes the emerging role of localism in America.

My thoughts on practical idealism came from Ken Jones's *Beyond*

Optimism: A Buddhist Political Ecology (Oxford, UK: Jon Carpenter Publishers, 1993).

The ideology of fatalism is discussed by Eileen Crist in "Beyond Climate Crisis: A Critique of Climate Change Discourse," *Tellos* 141 (2007). One proponent of fatalism is James Lovelock in his books *The Revenge of Gaia: Earth's Climate in Crisis and the Fate of Humanity* (New York: Basic Books, 2006) and *The Vanishing Face of Gaia: A Final Warning* (New York: Basic Books, 2009).

See Charles Hall and John Day, "Revisiting the Limits to Growth after Peak Oil," *American Scientist,* May–June 2009.

The Emancipation Proclamation, the document that abolished slavery in America, was signed in 1863.

My thinking on values and worldviews in a time of climate change was based on Tom Gallagher, "The Values Orientation Method: A Tool to Understand the Differences in Cultural Differences," *Journal of Extension* 39, no. 6 (2001), and Tim Kasser's seminar Human Identity and Environmental Challenges, presented at the Gund Institute at the University of Vermont on March 30, 2010, http://vimeo.com/10635948.

My figures on poverty in the world were taken from the World Bank Poverty Reduction and Equity Statistics and Indicators (2005), www.web.worldbank.org.

Derrick Jensen writes provocatively about the role of individual energy conservation in "Forget Shorter Showers," *Orion* magazine, July/August 2009, and again in "Side with the Living," *Orion* magazine, September/October 2009.

Author's Note on Cover Art

A *capriccio* is a lively, whimsical interlude in a classical music composition. An Italian word, it translates along the lines of "following one's fancy" or, literally, "caprice." One of the most famous composers of the style was Niccolò Paganini (1782–1840). His difficult, mischievous capriccios are filled with rapid trills, slurred staccatos, and places in which four-stringed violins mimic flutes. The most accomplished player quakes at his scores. Nevertheless, through the years musicians have adapted Paganini's capriccios so often that "variation on a theme by Paganini" has almost become a cliché. It is a measure of his genius that this music refuses to die; it only persists, continuously adapted.

But it is not only composers who employ the capriccio. Visual artists use the form when they improvise fantasy in the midst of structure or juxtapose architectural drawings with fictional or fantastical figures. Francisco José de Goya did just this when he drew a beautiful woman with the wings of an eagle flying above three calculating crones, or a priest with a parrot's head genuflecting in front of a crowd of mules.

The capriccios by contemporary artist Gail Boyajian

are full of whimsy. But they are purposeful, too, exploring the potential of America's landscapes to change as the climate warms. In *High Meadow Capriccio,* one of four paintings in a series Boyajian was commissioned by the Art of Action, a project of the Vermont Arts Council, to paint, cows and bison graze incongruously alongside ostriches and alpacas. In the distance, wind machines dominate the tops of mountains, while below, a high-speed bullet train barrels across a pastoral landscape and a man, driving a team of Morgan horses, races alongside. Blue flag irises, a wetland wildflower I've known since I was a child, are painted in the foreground, and next to them is a least bittern, executed with Audubon-like detail. Nearby, a lime-green grasshopper, an insect that evolved four hundred million years ago, surveys the setting while a dinosaur emerges from the woods.

Rather than exploiting the contrast between architecture and human figures as other capriccio painters have done, Boyajian juxtaposes an emerging technological infrastructure with living characters that occupy the landscape across time—past, present, and future. "Which epoch is this?" the observer is compelled to ask. All of them, all at once, it seems. Boyajian's paintings have us confront how innovation—that unique feature of human evolution—compares to the long-standing persistence of life. The ingenuity that led to the invention of bus-sized wind turbines contrasts with the creativity that begot the Blackburnian warbler, a bird the size of a child's fist that is looking in from the edge of the painting, having just flown five thousand miles from the highlands of Peru.

"In my paintings I work almost intuitively, elaborating on the small spaces," Boyajian tells me when we talk about her paintings for the first time. I lean closer and see the ecological niches she has worked to define. There is the flowering white mustard plant called cardamine, growing alongside a delicately painted hay-scented fern providing shade for a trout stream where a man fishes. There's the ruby-throated hummingbird nectaring from the thistle and the black-capped chickadee loosely suspended on the end of a pine branch. Each species—the mustard, the hummingbird, the man, and the trout—exists in a niche, a space where evolution has optimized its fitness. Each has adapted to the local conditions that it has found itself in.

Boyajian's *High Meadow Capriccio* is hung alongside *Village Center, Woods,* and *Valley Farmland,* landscape paintings that all collapse time and character into a single imaginary moment. Robert Frost contemplates the woods from a stone wall while Rachel Carson sights a warbler in the treetops. Calvin Coolidge fishes from the trout stream. The nineteenth-century towns in the background, complete with white steeple churches and apple orchards in their town centers, are holding out as twenty-first-century industry hovers on the fringe. All the while, magenta-petaled thistles and brown-eyed Susans witness human pursuits in the manner of a Greek chorus.

Boyajian's capriccios are speculations on the future. She acknowledges the agitation at hand while visually describing how things could unfold. "I'm an optimist," she tells me. "We have before us an opportunity to adapt, and I find it very stimulating." But the opportunity, Boyajian concedes, is a complicated one. "Extinction will be a part of the future as well," she says. In other words, the opportunity for adaptation comes with the sober reality of suffering.